EXPERIMENTAL PHILOSOPHY AND THE BIRTH OF EMPIRICAL SCIENCE

For my parents Lili and Dov, who offered me the truth of love

Experimental Philosophy and the Birth of Empirical Science

Boyle, Locke, and Newton

MICHAEL BEN-CHAIM

ASHGATE

Published by
Ashgate Publishing Limited
Gower House
Croft Road
Aldershot
Hampshire GU11 3HR
England

Ashgate Publishing Company
Suite 420
101 Cherry Street
Burlington, VT 05401-4405
USA

Ashgate website: http//www.ashgate.com

British Library Cataloguing in Publication Data
Ben-Chaim, Michael
Experimental philosophy and the birth of empirical science: Boyle, Locke and Newton
1.Newton, Sir Isaac, 1642-1727 2.Locke John, 1632-1704
3.Boyle Robert, 1627-1691 4.Science - History - 17th century 5.Science - History - 18th century 6.Science - Philosophy - History - 17th century
7.Science - Philosophy - History - 18th century
I.Title
509'.032

Library of Congress Cataloguing-in-Publication Data
Ben-Chaim, Michael.
Experimental philosophy and the birth of empirical science : Boyle, Locke, and Newton / Michael Ben-Chaim.
p. cm.
Includes bibliographical references and index.
ISBN 0 7546 4091 4 (hardcover : alk. paper)
1. Science--Methodology--History--17th century. 2. Science--Philosophy--History--17th century. I Title.

Q174.8.B46 2004
501'.09'032--dc22

2003025788

ISBN 0 7546 4091 4

Printed and bound in Great Britain by MPG Books Ltd, Bodmin, Cornwall

Contents

List of Figures

Acknowledgments

The publication of this book is a special opportunity to express my long-standing debt and gratitude to mentors, colleagues, and students who taught me to appreciate the challenge of understanding problems. My studies with Shmuel Eisenstadt and Joseph Ben-David at the Hebrew University, and Simon Schaffer and Nick Jardine at Cambridge University, were extremely valuable. I have also greatly benefited from learning and teaching in the stimulating environment of the Technion – Israel Institute of Technology.

The research for this book was supported by funds from the Technion. It was written during my term as a Visiting Scholar at the Department of the History of Science, Harvard University. I am grateful to Mario Biagioli and other members of the department for their caring hospitality. Nick Jardine and anonymous referees offered very helpful comments on an earlier draft. I thank Tom Yule, Antonia Aphantitis, and the editors of Ashgate for their assistance with preparing the manuscript for publication.

Chapter One

Introduction: the Historical Sociology of Scientific Explanations

The Historical Problem

Current research in the natural sciences is predominantly empirical. Minor and major advancements often involve the discovery of new materials and processes. Discoveries enable scientists to systematically observe and reproduce otherwise indiscernible effects of theoretical and practical value. Concepts and theories are rigorously defined by means of sophisticated observations and experiments. Procedures of empirical research are routinely applied in assessing the value of concepts and theories as tools in basic and applied research. For these reasons, in many research fields empirical work provides the necessary setting for the exchange of information between research workers. It is also widely considered to be an indispensable context of scientific education and training.

The empirical character of the natural sciences implies, in addition, that their gradual advancement is accompanied by a growing division of technical labor. An individual researcher is rarely able to master the technical knowledge necessary for carrying out an entire research project. This division of labor includes the industrial manufacture and commercial supply of research materials and instruments, prerequisites for routine work. Their cost often consumes the lion's share of the budget for research projects. The technological, intellectual, economic, and social aspects of empirical research are so closely related that together they form the culture of empirical research within the sciences.

This culture of empirical research, however, is a relatively recent phenomenon in the history of science. The philosophical aspects of empirical evidence and the role of experience in the production of knowledge had indeed already been studied in ancient Greece. These topics later became very common in the natural philosophical literature of the late Middle Ages. The philosophical literature included rudimentary reports of experimental studies which gradually expanded in scope and complexity throughout the Middle Ages, though most notably since the Renaissance. Yet the scope, complexity, and achievements of empirical research during these periods – in the occident or any other civilization – boldly pale in comparison to the modern culture of empirical research.

A valuable indicator of this historical and cultural difference is evident in early modern studies on electricity, as will be further clarified in the next chapter. Suffice it to note here that the scope and complexity of empirical research on electricity increased more during the second half the 18th century than it did throughout the entire preceding period of some two millennia, since Plato's discussion of electrical attraction in his *Timaeus*! In the 19th century, electrical research yielded unprecedented technological innovations that eventually transformed entire cultures of the human species despite the availability of similar research materials and instruments for centuries. It appears, therefore, that during the swift growth of experimental research on electricity in the 18th century, available materials and instruments were used by researchers in highly novel ways, as well as far more intensively and extensively than ever before. These differences indicate the birth of a radically new conception of the scientific meaning and value of empirical research.

This book explores the radical transformation that led to the emergence of modern cultures of empirical research in the natural sciences: How did it begin? Why did it happen? Who were its principal advocates? How were they related to one another? How did they manage to make the differences that eventually gave rise to a new culture of empirical research? Satisfactory answers to these questions may require examination of every field of empirical research and the cultural context in which it developed during the early modern period. This book presents a study within a considerably narrower scope, focusing on the works of Robert Boyle, John Locke, and Isaac Newton – the most notable progenitors of 17th century 'experimental philosophy'. This book's principal aim is to examine the impact of their experimental philosophy on the rise of the modern culture of empirical research.

Historiographic Considerations

Boyle, Newton, and Locke need not be introduced to scholars of science. This trio of English intellectuals has always figured prominently in studies related to the rise of modern science. The amounts of published literature on their lives and works exceed those on any other trio of scientists or philosophers so closely associated in history and geography. Though working through the wealth of primary and secondary sources relating their careers constitutes only one aspect of the research project presented in this book. The other aspect consists of a critical examination of historiographic and theoretical issues of import for science studies in general, namely presuppositions about empirical research as a means of doing science, and the aim of science itself. As will be shown in what follows, these issues are directly related to principal historical questions which form the focal point of this book.

Science studies have long been conducted within a disciplinary framework based to an important extent on two fundamental assumptions. The first assumption states that gathering and processing empirical evidence is a critical component of knowledge production. The second states that the aim of science is knowledge. These two assumptions have become so deeply and widely entrenched in science studies that they are now routinely taken for granted. Together, they form what may conveniently be labeled as the 'central dogma' of science studies, the dogma of 'science as knowledge'.[1] Debates about specific methodological or social definitions of knowledge and their historiographic implications have always formed an important part of the development of research in history, philosophy, and sociology of science. This book questions the adequacy of the central dogma itself as a frame of reference for reaching appropriate answers to the historical questions posed above.

By defining the aim of science in terms of an a-historical conception of knowledge, the central dogma offers a much too uniform image of the role of empirical research in science. If the aim of science – as

---

[1] I borrow the phrase, 'central dogma', from molecular biology, in the context of which it was coined by Francis Crick in the mid 1950s to denote the basic theoretical framework for studies on the flow of genetic information in living cells. I use it in the context of science studies analogically to denote a fundamental set of assumptions that is widely accepted by researchers and forms part of the definition of their common intellectual enterprise. Central dogmas, not unlike Thomas Kuhn's paradigms, mark out – intellectually and socially – the disciplinary core of research projects. The phrase is by no means applied in what follows to devalue the achievements of historical research, upon which my own study inevitably and indelibly rests.

the dogma claims – is knowledge based on experience, the culture of empirical research that currently characterizes the sciences should have gradually evolved since the era of ancient Greek philosophy when the methodological issues concerning the relation between theory and experience were thoroughly discussed. It could be expected, for example, that late medieval natural philosophers would routinely discuss the uses of experimental instruments in research and their value for training their students, or the intellectual and social aspects of the employment of technicians. Yet these issues are hardly mentioned in the literature from that period. If the central dogma was correct, it could be expected, more specifically, that the progress of experimental research on electricity during the second half of the 18th century would have gradually been achieved previously over the span of several centuries. Yet this gradual evolution clearly did not happen.

If the aim of science was always knowledge, and if it was always clear to scientists that knowledge hinges on empirical research, why do modern scientists invest many more resources in empirical research than late medieval scientists did? Why is empirical research considered more vital to the advancement of science in the modern era than ever before? It appears that modern scientists conceive of the uses and values of empirical research in a different way than their predecessors did, despite their shared methodological view that knowledge must be based on experience. Yet the central dogma, which considers empirical adequacy as the universal mark of knowledge, fails to suggest what could lead scientists to develop and value empirical research in such apparently radically different ways.

Historical studies on the works of Boyle, Locke, and Newton have been conducted predominantly within the historiographic framework of the Scientific Revolution, which has been specially established to address historical questions concerning the rise of modern science.[2] In accordance with the central dogma of science studies, however, historians have commonly assumed that the principal aim of empirical research in the natural sciences has always been the acquisition of knowledge. While drawing special attention to the importance of experience and experiment in the works of Boyle, Locke, and Newton, studies have been predicated on the assumption that the aim of empirical research did not change during the

---

[2] For expositions and surveys of studies on the Scientific Revolution, see Lindberg and Westman 1990; Cunningham and Williams 1993; Porter and Teich 1992; Shapin 1996; Henry 1997; Osler 2000.

Scientific Revolution. For this reason, historians have not systematically examined whether or not their experimental philosophy changed the aim of empirical research.

As will be shown throughout this book, the assumption that the views of Boyle, Locke, and Newton on empirical research primarily addressed the problem of knowledge, led historians to underestimate the novelty of their common endeavor to redraw the relationship between the use of experiment on the one hand, and the aim of philosophy on the other.

Suffice it to note at this stage some discrepancies between the historiographic assumption that experimental philosophers considered experiment to be a solution to the problem of knowledge on the one hand, and the evidence of their contributions to experimental research on the other. Historians have demonstrated, for example, that the common view among 17th century natural philosophers was that theory was inherently fallible and could at best be consistent with empirical evidence. Boyle clearly accepted these methodological maxims, but this in itself does not explain why he carried out experimental research projects on such unprecedented scales. His attitude towards experiment was often ascribed by historians as his adherence to Bacon's philosophy. However, Bacon stressed the importance of integrating experimental research with the systematic explication of the essential material qualities of physical bodies, while Boyle notoriously depreciated the philosophical value of such explications. Could Boyle simply adopt the experimental portion of Bacon's philosophy? Why would he do that when philosophers commonly taught that empirical evidence was valuable in as much as it was conducive to solving philosophical problems?

Newton seems to have gone further than Boyle in understating the importance of philosophical explications of the nature of things. Historians have associated his methodological views with his celebrated assertion, 'hypotheses non fingo'. Newton, however, also adhered to the fallibilist image of knowledge. It therefore appears that this image cannot explain his attitude toward hypotheses that were as consistent as possible with empirical evidence.

Consider, more specifically, Newton's research in optics. His discovery of the heterogeneous composition of sun light – the outcome of his most elaborate experimental research project – was fallible, as its early critics repeatedly stressed. His disciples claimed that it nevertheless offered a new model of experimental research in natural philosophy. But if it was as fallible as the theories of his predecessors – and for the same epistemological reasons – its impact on the history of science would not be directly related to the problem of the empirical foundation of knowledge. Historians have often attributed this discovery to his exceptional abilities; yet allusions to his talents do not explain what his special skills as an experimental investigator precisely modeled, or how innovative the model was.

Unlike Boyle or Newton, Locke did not specialize in natural philosophy. His *Essay concerning Human Understanding* has nevertheless been considered a milestone in the history of philosophy. Working their way from the central dogma of science studies, commentators have assumed that the importance of his *Essay* primarily pertained to Locke's 'theory of knowledge'. It has been demonstrated, however, that the principal tenets of the theory – namely that experience is the source of knowledge, and that the search for knowledge about the natural world does not transcend the scope of empirically reliable theories – were widely accepted by philosophers long before Locke began to write drafts for his *Essay*. It is therefore highly unlikely that Locke considered the theory of knowledge as the principal achievement of his *Essay*. His book was apparently important despite the conventional theory of knowledge it espoused.

Despite of the continuous growth of historical research on Boyle, Newton, and Locke, historians have paid relatively little attention to their common *and* distinctive interest in empirical research and its role in science. From the perspective on science as knowledge, their experimental philosophy appears to have merely affirmed the common methodological view that knowledge of the natural world hinged on accurate empirical research, and that experiment could provide a highly effective means to achieve this goal. It therefore appears that the dogma of science as knowledge fails to provide an adequate framework for understanding the special impact of experimental philosophy on the history of science.

This book suggests, accordingly, that studying the aim of empirical research in the experimental philosophy of Boyle, Locke, and Newton in particular, and in science in general, ought to be predicated on a revision of the central dogma. Knowledge about the world clearly presupposes empirical research. But is knowledge the aim of science? An alternative answer comes to mind once the central dogma itself is considered from a broader historical perspective.

The central dogma originated in positivist doctrines that became intellectually fashionable in the late 19th century, then dominated academia in the early part of the 20th century, and, somewhat ironically, have been in decline ever since. The positivists regarded the progress of science as an achievement that ought to command the respect of any reasonable human being, irrespective of his or her cultural identity. They accordingly sought to demonstrate that the achievements of science could be appraised solely by reason and experience, which are apparently the most universal and reliable properties of human cognition.

The image of science as a pursuit of knowledge based on human reason and experience gave credence to the ideal of science as a universal achievement untainted by historical and cultural specificities. Yet this ideal itself has been considerably undermined in recent decades. Science studies have greatly enhanced new modes of appraising science as a cultural tradition. Should we not opt out of the central dogma of science as knowledge?![3]

Before the rise of positivism, science was commonly associated with the desire to understand. The philosophical study of the natural world in particular was traditionally construed as being an intellectual endeavor with an aim of causal explanation of natural phenomena. The positivists did not necessarily renounce a desire to understand. They clearly recognized, however, that it could not possibly be discharged in an objective or culturally independent manner. Explanations are inherently pragmatic. Their meaning and value do not completely transcend the cultural idiosyncrasies of those who propose or accept them.[4] For this reason, the positivists understated the relevance of causal explanations to science. They accordingly confined the scope of knowledge to logically and empirically reliable propositions, and claimed that this knowledge was the aim of science.

A highly economic search for an alternative to the central dogma of science as knowledge would, therefore, be predicated on reinstating the traditional notion that the aim of science is causal explanation. As will be clarified in what follows, cognitive reliability is clearly a critical quality of explanations, and for this reason, the problems of knowledge and explanation cannot be completely separated from one another. The image

---

3 For pioneering studies that question the value of the image of science as knowledge, see, for example, Taylor 1987; Jardine 1991; Pickering 1995, 1995a.

4 For a discussion of the pragmatic characteristics of explanation in contrast with the non-pragmatic status of scientific knowledge, see van Fraassen 1980.

of science as the pursuit of causal explanation nevertheless provides a new perspective on empirical research in science and its history.

Scientific Explanations and their Sociological Study

From the perspective on science as knowledge, empirical research has been construed as a means to obtain evidence. It has been claimed that the function of empirical evidence was methodological, and could be explicated in terms of inferences that determined whether theories could withstand the test of evidence.

The perspective on science as causal explanation suggests a different function for empirical research. Causal explanations are creative endeavors that aim to render experience more intelligible and thereby satisfy the desire to understand. As Hume pointed out, because they convey beliefs concerning the causes underlying empirical evidence, they can neither be refuted nor proven by it. Yet empirical evidence is by no means irrelevant to the desire to understand. To be cognitively reliable, explanations must account for experience in a logically consistent manner.

However, in virtue of their pragmatic qualities, explanations may account for experience in different ways. In this respect, the creative endeavor to understand may conveniently be construed in terms of other forms of human creativity, namely the transformation of resources into goods. Empirical information is the resource of which explanations are partially made and rendered cognitively reliable. The value of empirical evidence is therefore a *variable* that depends on pragmatic considerations. Endeavors to satisfy the desire to understand may accordingly assign different values to empirical research. Thus, from the perspective on science as causal explanation, it is not surprising that scientists in modern society relate empirical evidence to their theories in ways that differ considerably from those that were devised by their forefathers under remarkably different social and cultural circumstances.

Despite the inherent variability of the relationship between explanation and empirical information, the very existence of traditions of training and research in the sciences indicates that scientists seek to regulate the production of explanations. Scientists do not merely seek to satisfy their own desire to understand. They search for explanations that are likely to benefit others and be accepted by them for this reason. In this respect, explanation is a social activity that is shaped by social values *and* intellectual activity shaped by intellectual values. The intellectual work that is involved in the transformation of empirical information into explanations is guided by social interest in their public transaction. Conversely, the effort invested in the public transaction of explanations is shaped by intellectual interests that are invested in their production.

Continuity and change in the transformation of empirical information into causal explanations involve, therefore, continuity and change in its social regulation. Understanding historical variations in the relationship between empirical information and causal explanation is accordingly a historiographic and sociological problem. Social historians commonly apply sociological tools to the study of social interactions and exchange that are integrated into forms of social structure and organization. From a post-positivist perspective that recognizes the sciences as being integrated with other institutions in society, the sociological concept of social institutions appears to be the obvious tool to employ in studying the variable function of empirical research in the history of science.

Social institutions are regulatory constructs that set constraints on, and offer opportunities for, the satisfaction of human needs and desires. They are comprised of explicit rules and implicit codes which individual agents learn to recognize and assimilate through processes of socialization. They are the fundamental social means for generating behavioral patterns across temporal and spatial singularities of human actions, and for rendering the subjective aspects of behavior into communicable idioms.[5]

In historical studies on science, institutions are often identified with formal organizations such as universities, national academies, or professional associations. Sociologists tend to broaden the definition of institutional activities to include social transactions that are regulated in more informal and implicit manners. Formal and informal institutions are often integrated with one another in shaping the various domains of human life and action. For example, scientists often conduct their research under auspices of universities or industrial firms; yet the social transaction of their research is institutionalized by means that transcend official regulations of those organizations.

The sociological perspective on science as the pursuit of causal explanations is based, more particularly, on the aforementioned assumption that although explanations are products of the creativity of individual scientists, the latter have a vested interest in regulating the social

---

5 North 1991.

transaction of their creative endeavors. This interest engenders the institutionalization of explanatory practices in general, and specifically those that transform empirical and intellectual resources into valuable explanations. Individual scientists contribute to the institutionalization of explanatory practices by shaping them into modes that are applicable by other researchers to the study of diverse problems. These modes function as regulatory devices by conveying specific constraints on, and opportunities for, the expansion of research.

From a sociological perspective, then, regulating the social transaction of scientific research is considered a problem that scientists themselves aim to solve. Different solutions give rise to different institutions. The comparative study of institutions provides, in turn, the framework for understanding different modes of transforming empirical resources into reliable explanations. Premised on the assumption that the intellectual and social interests of scientists are complementary, this study stands squarely against positivist attempts to explicate scientific research in terms of universal methods incarnated in the rationality of individual scientists.

Our sociological study differs, moreover, from anti-positivist portrayals of science as a creative endeavor that must be subdued by the dictates of social conventions or the pressure of certain groups. Critics of positivism have assumed that if science cannot be regulated by a universal method, its public credentials must be manifestations of social compliance. According to Kripke, for example, Wittgenstein's 'later' philosophy has demonstrated that the 'successful' performance of cognitive tasks 'depends on the brute fact that we agree with each other in our responses'.[6] Kuhn has similarly suggested that the dogmatic adherence to 'paradigmatic' solutions to research problems is the necessary condition for scientific growth.[7] According to Latour, the 'first principle' of the social study of science is that 'the fate of facts and machines is in later users' hands; their qualities are thus a consequence, not a cause, of a collective action'.[8]

---

6 Kripke 1982, p. 109, with special reference to Wittgenstein's studies of rules and private language in Wittgenstein 1976. For an application of Wittgenstein's philosophical ideas to laboratory studies, see Bloor 1983; 1992; Lynch 1984, 1992.

7 Kuhn 1962, 1963. For studies that expand upon the role of conventions in science, see Barnes 1974; Collins 1985.

8 Latour 1987, p. 259.

Ironically, the role of modern empirical science as a catalyst of social change undermines the plausibility of conventionalist accounts of social order. The image of science as knowledge implies, however, a dichotomy between the cognitive reliability of human judgment on the one hand, and its credibility as a social convention on the other. Given this dichotomy, collective actions appear to be the only conceivable alternatives to the impersonal methodological evaluation of claims to knowledge. The sociological study of scientific explanations dissipates this dichotomy. Rather than proposing a *new* alternative to positivism, it reaffirms the age old image of science as an endeavor to advance human understanding. Assuming that causal explanation is a principal means to achieve this goal, sociological research aims to explain how intellectual and social interests reinforce one another through the efforts of researchers to institutionalize their explanatory practices.

The perspective on scientific explanations as social institutions does not imply that scientists should conduct their research in socially harmonious ways which they never seek to change. Quite to the contrary, the constraints and opportunities that regulate the production or transaction of explanations are clearly not immutable. They may become, at any time, the subject of reappraisal and change. From a sociological perspective, the history of the human species represents continuous institutional changes. Within this broad context, the study presented in this book considers the history of science as the succession of institutional changes in explanatory practices. More specifically, it sets out to examine the contributions of Boyle, Locke, and Newton to the institutional change of modes of explanation that engendered the development of a new culture of empirical research in natural science.

The Main Thesis of the Book and its Narrative Structure

This book focuses on the relationship between empirical research and especially experimental practices on the one hand, and explanatory practices on the other. It aims to demonstrate how the experimental philosophy of Boyle, Locke, and Newton transformed this relationship. Moreover, it aims to show how the mode of explanation that became the hallmark of experimental philosophy engendered more robust empirical research than ever before in the history of western philosophy.

Before Boyle, the oldest of the trio, philosophers commonly sought causal explanations that were predicated on the definition of the nature, or essence, of things. This mode of explanation was traditionally acknowledge as the legacy of ancient Greek philosophy. The main goal of the academic discipline of natural philosophy was to provide students with the proper tools to seek definitions of particular natures and elucidate their explanatory value.

Following in Boyle's footsteps, Locke and Newton established models of a new mode that aimed to explain how empirical properties of physical bodies were correlated. This mode fell short with respect to disclosing particular natures, though it nevertheless posed new challenges and promised novel achievements. Understanding empirical correlations hinged on the methodical cultivation of practical experience. To discover how empirical properties were correlated, philosophers had to develop conceptual tools and techniques to discern and predict correlations and, when possible, to reproduce them by experiment. The new mode of explanation was advocated as an 'experimental philosophy' because it posed the philosophical problem of explanation in a manner that rendered empirical research vital to its solution.

As will be shown throughout this book, the development of the experimental mode of philosophical explanation did not involve substantial revisions to prevailing theories of knowledge. Methodological procedures such as induction and deduction, or analysis and synthesis, were adapted and applied to the new models of causal explanation. For this reason, historians who followed the central dogma and focused their studies on the relationship between empirical evidence and claims to knowledge have failed to adequately discern how innovative experimental philosophy was.

However, the transition from the traditional philosophical mode of explanation to this new one effected a major cultural change because the two modes were associated with different values of human learning, different models of human thought and cognition, different research practices, and different forms of the social transaction of their results. The narrative of the book is accordingly structured to elucidate the process of institutional change in explanatory practices and its many cultural layers.

Institutional changes evolve throughout historical time, and always engage the efforts of more than one person. Moreover, they are neither masterminded nor foreseen by any single person who partakes in bringing them about. It has therefore appeared helpful to conceive of and narrate this institutional change by means of a metaphor of a 'train of thought' that is conducted by different individuals, and gradually changes its course in a manner that eventually leads to a completely new destination. While the initial deliberation of change may not seem decisive, its final phase is relatively self-evident.

The clearest and most coherent historical evidence for the transition from old to new modes of explanation pertains to the work of researchers who took it for granted and were barely cognizant of the older tradition, its ideals, and the practices that characterized it. Historical evidence for the initiation of the transition, by contrast, is considerably more ambiguous. It indicates primarily programmatic attempts to envisage and conceive of new alternatives in contexts still thoroughly dominated by the prevailing tradition.

To clearly demonstrate this institutional change, the narrative of this book has been structured in a way that is admittedly uncommon to historical research. It begins with an examination of the radical break with the traditional philosophical discourse on the nature of things, and then gradually proceeds backwards in chronological order to examine how the transition was initiated.

The earliest unambiguous evidence for the new mode of causal explanation has been found in experimental studies on electrical conductivity from the early part of the 18th century. These were initiated by Stephen Gray, an amateur naturalist associated with the Royal Society. **Chapter Two** examines how the discovery of electrical conductivity brought a tradition of philosophical reflections on the nature of electric bodies with origins traceable to the writings of Plato to an end. Focusing on the public transaction of the uses of electrical conductivity as an explanatory construct and an experimental technique, this chapter further examines how their institutionalization brought about unprecedented developments in electrical research. To identify the historical origin of Gray's novel approaches to the explanation of electrical phenomena, this chapter further examines Gray's early career, especially both his direct and indirect exchanges with Newton's assistants at the Royal Society.

By the turn of the 18th century, Newton's experimental philosophy was commonly associated with a new theory of light and colors, disclosed to the public in his first published work in an article from 1672. **Chapter Three** focuses on Newton's early studies in optics, when he was a student at Trinity College in Cambridge. These studies began with the methodical assimilation of philosophical works concerning the nature of light and colors, especially Descartes' work. They further indicate, however, his strategic decision to redesign his entire approach to the study of light and colors. The chapter examines how Newton discarded his Cartesian premises, and instead began to develop experimental techniques for the analysis of the formation of spectral colors. His published article centered around a concise factual account of this analysis that conveyed at the same time his explanation of the origin of colors.

In this publication, Newton presented his experimental solution to the problem of the origin of the spectrum as a turning point in the history of the philosophical discourse on light and colors. **Chapter Four** demonstrates how his article was rhetorically structured to convey the new role experimental work could play in the transaction of causal explanations. Historical examination of Newton's rhetoric shows its similarity to the biblical genre of sermonical preaching, which was originally developed by Protestant critics of the philosophical style of scholastic preachers.

Newton's audience included scholars who were already reputed for their contributions to natural philosophy, and refused to acknowledge the authority of his experimental analysis of colors. As shown in **Chapter Four**, the controversy between Newton and his critics was centered around their clashing views on the relationship between empirical research and causal explanation in natural philosophy. In the midst of this controversy, Newton set out to explicate the rules for demarcating his 'experimental philosophy' from the philosophical discourse on the nature of things.

How Newtonian was Newton's experimental philosophy? The next two chapters suggest that the answer to this question directly pertains to the philosophical studies of his contemporary and compatriot Locke. While Newton searched for a factual account of the origin of spectral colors, Locke was engaged in constructing a general model of human understanding in which the search for causal explanations was restricted to the representation of empirical evidence.

With the rise of positivism, Locke's *Essay* became the target of critics who sought to demarcate the 'theory of knowledge' from quasi-psychological studies of human cognition. **Chapter Five** shows how Locke and his contemporaries, by contrast, were primarily concerned with the problem of explanation. They accordingly construed the study of human cognition and thought to explain how human beings could methodically cultivate philosophical reasoning about natural order.

Philosophers traditionally taught that images of material objects were transmitted from the sense organs to the intellectual faculties. These transmitted images could then be abstracted by methodical analysis that ascertained more universal properties of particular objects. **Chapter Five** shows how Locke utilized the received view that experience was the basis for understanding to radically reverse the traditional model of intellectual processes. Locke's new model hinged on his demonstration that images of material objects were not sensory imprints, but rather intellectual constructs that were designed to represent empirical impressions in particular ways. These representations fell short of disclosing the nature of things; yet they could nevertheless be methodically improved as a means of understanding how empirical qualities were correlated in one way rather than another.

Chapter Six further examines Locke's new model of human understanding in light of his moral and political views concerning the institution of natural law. Locke's endeavor to bring these views to bear on his study of human understanding led him to reexamine the relationships between knowledge, faith, and human action. This context reveals the breadth of his social and historical perspectives on experimental philosophy and its leading role in binding faithful conduct and secular cultivation of the public good.

Newton and Locke committed themselves to advancing experimental philosophy in the late 1660s, long before they became acquainted with one another. They both contributed to the consolidation of a new mode of explanation in ways that were highly complementary. This historical coincidence is explained in **Chapter Seven,** which traces their common cause back to Boyle's programmatic writings on experimental philosophy as a vocation. In a pioneering study on Boyle's experimental philosophy, Shapin and Schaffer have demonstrated that Boyle consistently sought to separate philosophical discourse on the nature of things from public transaction of matters of fact.[9] The present study sets out to show how Boyle conceived of statements of fact as media for transacting a new kind of causal explanation.

Unlike Locke or Newton, Boyle was a prolific author and devoted none of his numerous publications to a comprehensive presentation of the principles of experimental philosophy. His work aimed to elucidate the different aspects of the calling of experimental philosophy in relation to a wide variety of topics on natural philosophy and practical sciences, as well as morality, religion, and theology. **Chapter Seven** examines how Boyle unified his diverse scholarly interests in his unremitting endeavor to redesign natural philosophy as a model of divine worship. Shaped by the

9 Shapin and Schaffer 1985.

values of Christian humanism, his conception of worship was predominantly secular, practical, and intellectual. This enabled him to conceive of empirical research in a new light as a critical tool for understanding God's worldly dominion.

Chapter Two

The Break with the Philosophical Discourse on Nature: The Discovery of Electrical Conductivity

Introduction

The 16th and 17th centuries bore witness to the invigorated attention of scholars to the empirical sources of knowledge. Notable factors gradually coincided to engender a stricter adherence to the evidence of the senses: the diffusion of practical information and reports through the printing press enabled scholars to systematically identify erroneous information with greater ease. Geographic explorations and discoveries dramatically disclosed discrepancies between genuine evidence and entrenched opinion. Travel within Europe and abroad became easier and more frequent, and scholars could more effectively check reports and stories that came by. The development of pedagogical instruments of literacy by humanist scholars upgraded the status of primary sources. The emphasis on literate piety in the Reformation and Counter-reformation movements highlighted the methodical attention to revealed sources of faith.[1]

Further impact on the methodical collation and processing of empirical evidence was made by the gradual rise in social and intellectual status of practically oriented disciplines such as the mixed-mathematical sciences of astronomy, optics and mechanics; related operative disciplines such as architecture and cartography; medical research; and the leisured pursuit of natural history in polite culture. The ascent of these branches of

1 Jones 1965; Debus 1978; Eisenstein 1980; Greenblatt 1991; Reeds 1991; Grafton 1992.

learning enhanced the value of measuring devices, instruments of observation and experimental research, and methods of classification.[2]

The overall lesson drawn from these developments was rhetorically manifested in reiterated references to the 'book of nature' as the only authority worthy of the unqualified respect of natural philosophers. As Shapin noted, 'no seventeenth-century modernist maxims seem more self-evidently sound than these: rely not on the testimony of humans but on the testimony of nature; favor things over words as sources of knowledge; prefer the evidence of your own eyes and your own reason to what others tell you'.[3]

In modern historical literature, the recurrent appeal to experience and reason in the early modern period has often been presented as a corroboration of the basic image of science as knowledge. In addition, it has suggested at least a partial explanation for the rise of modern science: whenever and wherever the cultural, social, or political circumstances were conducive to the relentless endorsement of reason and experience as the ultimate sources of knowledge, learning advanced beyond old, untested ideas. The path toward modern science could thereby become more secured.

The advancement of the study of electrical phenomena has been considered to be an outstanding illustration of this path. Unlike disciplines that had already been well established by the late Middle Ages – such as astronomy, mechanics and optics – electrical phenomena were considered a minor topic of philosophical research in the 17th century. Historians have accordingly assumed that electrical research did not have to be 'revolutionized' to become modern. Instead, it could gradually grow and flourish with the support of the new attitudes towards reason and experience. 'The study of electricity', as Hackmann noted, 'was in its infancy', 'everyone with the appropriate apparatus could make discoveries and contribute to its development'.[4] Upon a solid basis of empirical discoveries, the early seeds of modern mathematical physics were safely planted by the quantifying physicist who 'accomplished his purpose', as

2 Zilsel 1942, 1945; Hacking 1975; Ashworth 1980; Westman 1980; Feingold 1984; Murdin 1985; Bennett 1986, 1991; Biagioli 1989; Van Helden 1983, 1995; Impey and MacGregor 1985; Haddon 1994; Findlen 1994; Dear 1995; Hankins and Silverman 1995.

3 Shapin 1996, p. 69.

4 Hackmann 1978, p. 3.

Heilbron noted, 'by dropping scruples that had weighed with his predecessors and by adopting an instrumentalism that would have scandalized most of them'.[5]

The growth of knowledge of electricity could attract, in turn, more researchers. According to Heilbron's unsurpassed historical survey, merely 27 investigators reported to have studied electrical phenomena throughout the 17th century, as compared to striking increase of 88 in the fourth decade of the 18th century.[6] With the social organization of these new recruits, the modern science of electricity has appeared – from the perspective of the history of science as knowledge – to be safely on its way.

In light of the first chapter's reappraisal of the central dogma of science studies, this second chapter sets out to question the historical adequacy of the image of experience and reason as the twin engines pulling along progress of electrical research. Was the culture of 18th century electrical research based on new knowledge, or was new knowledge a product of institutional changes in the use of reason and experience? Did discoveries of phenomena challenge old philosophical explanations, or were the latter displaced by new ideals and modes of making sense of empirical evidence? Did the growth of electrical research vindicate accepted methods of empirical knowledge, or was it predicated on a transformation of the relationship between empirical and explanatory practices?

These questions are examined in detail in what follows in reference to one of the cornerstones of the new science of electricity, namely the discovery of electrical conductivity in the early part of the 18th century. The first two sections are structured as a comparative study, and for this reason the narrative style in the exposition is relatively understated. The first section focuses on explanations of electrical phenomena that were proposed by two philosophers who were renowned for their contributions to modern natural philosophy – William Gilbert and René Descartes. The second section addresses Stephen Gray's original account of electrical conductivity from 1731, and examines how different his explanation of electrical phenomena was. The third section examines the institutional change in electrical research that followed Gray's innovative work. The fourth section addresses the historical and cultural context of Gray's

5 Heilbron 1979, p. 1.
6 Ibid., pp. 99-100.

innovative work, in an attempt to identify the principal agents that gave direction to his creative endeavors.

Philosophers and the Nature of the 'Electron'

Electrical phenomena before the 17th century were predominantly associated with the motion of light objects in the presence of the 'electron', which was the Greek word for amber. Until the 18th century, the study of these phenomena often digressed from philosophers' main themes of inquiry. This is indicated in part by the relative volume of work devoted to the subject. For example, in Gilbert's *On the Magnet*, which, by the late 18th century, was commonly acknowledged to have inaugurated the modern science of electricity, the discussion of electrical phenomena was confined to only one of the book's 115 chapters. Descartes presented his philosophical examination of the subject in a couple of paragraphs in his *Principles of Philosophy,* his longest and most ambitious work on natural philosophy.

Published in 1600, *On the Magnet* was a contribution to the new heliocentric cosmology that was previously advocated by Copernicus. The plausibility of this new cosmology hinged in part on the explication of the planetary motions of the earth. Gilbert's book was primarily concerned with this problem, and one of its main goals was to demonstrate that circular motion 'is innate in the terrestrial globe, and natural to it', in virtue of its magnetic properties. His research project was inspired by the magnetic compass, unknown to ancient scholars who often associated loadstone with amber. Gilbert thus sought to demonstrate that the ancients did not adequately distinguish between these two minerals. In this context he set out to carefully examine the properties of the electron.[7]

Common classifications associated amber with loadstone in virtue of their shared and apparently exceptional property to attract surrounding light objects. In virtue of their common nature as minerals, however, philosophers traditionally reasoned that the explanation for electric and magnetic phenomena must be predicated, in part, on a definition of the nature of all minerals. This reasoning was based on the assumption that physical phenomena were generated by physical bodies, and that the former must be explained, accordingly, by the proper definition of the essential

---

7 Kay 1981, esp. pp. 82-5; Freudenthal 1983.

properties of the latter. The search for such a definition gave rise, in turn, to systemic philosophical explanations that ultimately encompassed the entire realm of particular natures and their interrelations.

Within this framework, Plato set the groundwork for subsequent debates about the nature of amber. As he explained in his cosmological treatise, *Timaeus*, the attraction to amber and to loadstone 'at which men wonder' was generated by imperceptible materials that linked the attractive and attracted bodies together. He accordingly stressed that 'there is no real attraction in any of these cases'. 'Proper investigation', he professed, 'will make it plain that there is no void; that the things in question thrust themselves around, one upon another; that the several kinds of body, as they are disintegrated or put together, all interchange the regions towards which they move; and that the results which seem magical are due to the compilation of these effects'.[8] Following Plato, ancient scholars such as Aristotle, Epicurus, Plutrach, Lucretius, and Galen suggested that amber and loadstone contained effluents that brought forth the motion of environing bodies.[9]

Gilbert conducted empirical studies which were unprecedented in their rigor and scope to elucidate the differences between magnetic and electrical attractions. He thus discovered that the latter, unlike the former, were generated by many different materials in addition to amber, which he labeled *electrics*. Notable among these materials were glass, which later became the standard electric in philosophical research, as well as diamond, sapphire, sulfur, and sealing wax. Careful observations demonstrated, moreover, that loadstone attracted only iron, but electrics were able to attract various materials so long as they were sufficiently light and in close proximity to the electric body.[10] This evidence enabled Gilbert to present his theories as novel contributions that advanced philosophy beyond 'former times when philosophy, still rude and uncultured, was involved in the murkiness of errors and ignorances'.[11]

It would, however, be highly anachronistic to construe Gilbert's empirical studies within a framework of modern theories of knowledge.

---

[8] Cornford 1997, pp. 319-26. Plato's comments on the subject were included in Sections 79e-80c, which formed an appendix to his discussion of terrestrial phenomena involving circular motion. See also Kay 1981, pp. 33-4.

[9] Kay 1981, pp. 31-46.

[10] Gilbert 1941, pp. 77-8.

[11] Ibid., p. 1.

Gilbert was engaged in a critical dialogue with his predecessors whose principal aim was to explicate the differences between the nature of loadstone and the nature of electrics.[12] The philosophical ideas he reviewed were often too abstract to be either conclusively refuted or affirmed by his own empirical findings. Empirical and philosophical resources nevertheless provided him with an arsenal of reasons with which he elaborated his own definition of the nature of electrics. Thus, following a critical review of Galen's classification of instances of natural attraction, Hippocrates' analysis of the relation between moisture and the growth of plants, as well as more recent theories of heat and its relation to local motion, Gilbert concluded that the nature of electrics could best be defined in terms of the composite nature of the earth. Minerals that originated from earthly 'humors', but 'possess sufficient firmness', were likely to generate an electric effect, especially upon attrition.[13] In accordance with the received philosophical view, Gilbert rejected the 'vulgar' notion of attraction and suggested that the electric body, in virtue of its humorous nature, generated the subtle effluent that drew the surrounding bodies toward the electric.

Gilbert's explanation of electrical phenomena illustrates the character of natural philosophical research as an organized contest of intellectual authority. Careful attention to empirical evidence as well as to broad philosophical considerations distinguished his arguments concerning the nature of electrics from lay opinion. These arguments concomitantly established his authority and challenged the authority of his predecessors. In theory, truth was the ultimate determinant of philosophical disagreements. As Descartes, for example, noted, 'whenever two persons make oppose judgments about the same thing, it is certain that at least one of them is mistaken…; for if the reasoning of one of them were certain and evident, he would be able to lay it before the other in such a way as eventually to convince his intellect as well'.[14] In practice, however, philosophers recognized that their common search for the nature of things offered each and every one of them the opportunity to dispute received views and to partake in the contest for learned authority. This is in part how their common pursuit could form organized cultural activity.

In the spirit of scholarly disputation, Descartes professed in his *Discourse on Method* (1637) that his 'plan has never gone beyond trying to

---

12 Kay 1981, 60-82.

13 Gilbert 1941, pp. 83-4.

14 Descartes 1985, i, p. 11.

reform my own thoughts and construct them upon a foundation which is all my own. If I am sufficiently pleased with my work to present you with this sample of it, this does not mean that I would advise anyone to imitate it. Those on whom God has bestowed more of his favours will perhaps have higher aims'.[15] His 'own thoughts' on the nature of electrics – not unlike Gilbert's – revealed his agreement with certain views and his disagreement with others. Like Gilbert, he followed the assumption that electrics contained an effluent material; but he rejected the peripatetic theory of elements that Gilbert considered sufficiently 'plain to all'.[16] Although Descartes admitted that a comprehensive explication of the nature of electrics required a thorough investigation of their mineralogical properties, he dissociated the study of electrics from the discourse on the specific nature of the earth. While Gilbert's interest in electrics stemmed from his broader endeavor to explicate the circular motion of the earth, Descartes considered the study of electrical phenomena as an opportunity to expand upon his conception of physical space.

According to Descartes, three fundamental grades or 'elements' of matter formed the universe of bodies in motion. The first was comprised of very small fast-moving particles of which the sun and the stars were made. The second was comprised of ether, which consisted of slightly larger, but still imperceptible, spherical particles that filled the physical universe. The third was comprised of larger particles of which terrestrial bodies were commonly made.[17] The typical electric, he explained, was a relatively porous material permeated by the second element of ether (see Figure 2.1).[18] Transparent solids such as glass were highly porous, and for this reason could be easily transformed into electrics. Descartes examined the properties of glass in other contexts; most notably in connection with his study on light and its refraction, and perhaps it was for this reason that he construed the nature of electrics in terms of the nature of glass. Although amber was admittedly more opaque, Descartes suggested that the humors from which it originated turned it into a relatively porous solid.

By the time Descartes examined electrical phenomena, a Jesuit scholar from Ferrara, Niccolò Cabeo, demonstrated that Gilbert's

---

[15] Ibid., p. 118.

[16] Gilbert 1941, pp. 83-4; for his peripatetic chemistry, see Freudenthal 1983. For Descartes' theory of matter, see Garber 1992, pp. 63-93, 117-36.

[17] Descartes 1983, p. 110.

[18] Ibid., pp. 238-9.

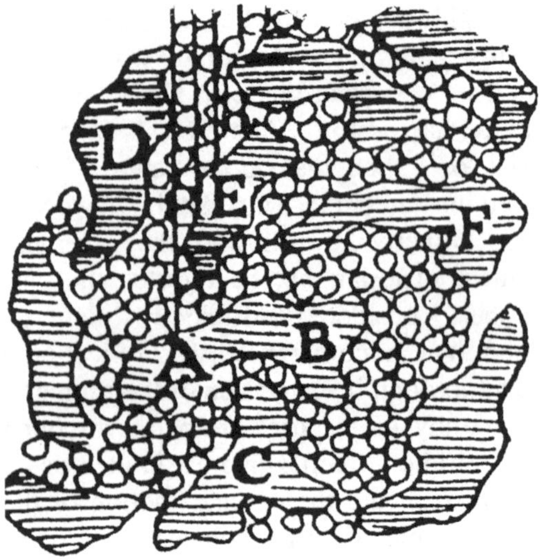

2.1 Descartes' illustration of a porous body (composed of parts A-F) permeated by the spherical particles of the 'second element' (Descartes 1677, by permission of the Houghton Library, Harvard University)

rebuttal of the theories of the ancients was premature. Following in their footsteps, Cabeo professed that electrical phenomena were the product of a vacuum effect. The attrition of electrics released subtle effluvia that rarefied the surrounding air, generating in turn a small whirlpool. This better explained, in his opinion, why the surrounding bodies were drawn to the electric but were also repelled away from it. Electrical attraction was by nature inseparable from electrical repulsion.[19]

Taking these factors into account, Descartes professed that the attrition of electrics agitated the spherical particles of the ether enclaves which then expanded into the surrounding air, pushing the surrounding bodies away from the electric. The expansion of ether caused electrical repulsion. However, since the surrounding air was structurally more homogeneous than porous electric, it blocked diffusion of ether particles and pushed them back into the electric, thus generating the vacuum effect of electrical attraction that Cabeo and his predecessors proposed.[20]

---

19 Heilbron 1979, pp. 181-3.

20 Ibid., pp. 273-4.

Electrical Conduction: An Explanation that Made a Difference

The debate amongst philosophical authorities on the nature of the electric body was perpetuated over two millennia. Studies throughout the 17th century demonstrated that this could continue even longer. In the early 1730s, however, the debate was brought to a halt and radically transformed as a result of the discovery of electrical conductivity. The discovery was originally announced by Stephen Gray, an amateur naturalist from Canterbury, in a brief experimental report to the secretary of the Royal Society, Sir Hans Sloane. Shortly thereafter, the report was published by the Society in its *Philosophical Transactions*. Gray's report apparently did not fit in with the traditional discourse on the nature of electrics, and the discord was striking. In 1732, Sloane wrote to his colleague, the abbé Bignon from the Académie Royale des Sciences in Paris, that 'these experiments on electricity [are] surprising, no one here is offering to explain them'.[21]

Gray's report did not address the philosophical problem of defining the nature of electrics; though it clearly offered a causal explanation for the familiar phenomena of electrical attraction and repulsion. The explanation, moreover, was the experimental account itself. Gray reported to have carried out an experimental project which at its core consisted of the transmission of familiar effects of attrition of an electric through various materials, which he termed 'lines of communication', that extended several yards away from the electric body. None of these lines, Gray stressed in his report, was itself an electric. The process of electrical 'communication' was nevertheless the cause of electrical effects: when the lines of communication were removed from the experimental setting, the phenomena failed to occur. When the lines were properly placed, the phenomena were predictably observable. Electrical communication was clearly a necessary condition for the production of electrical phenomena. The explanation, however, effaced the fundamental distinction between phenomena and the nature of things which traditionally structured the context in which philosophers plied their trade. Electrical communication represented neither a specific phenomenon nor a particular nature.

---

21 Quoted from Heilbron 1979, p. 252.

Gray's experimental report did not explicitly undermine the value or legitimacy of the philosophical discourse on the nature of electrics. It nevertheless suggested precisely how irrelevant the discourse had become. Philosophers traditionally based their studies on the conviction that electrics were by nature endowed with a special quality. The conviction, as noted earlier, followed the assumption that every physical body had a particular nature which determined the principal observable properties of that body. Gray, by contrast, distinguished between the properties of the electric body and the principal cause for the electrical phenomena. He of course recognized that the presence of an electric body was necessary. At the onset of his report, however, he introduced the term 'electric vertue' to denote a physical property that could be attributed to electrics as well as to other bodies, noting that 'the Electrical Vertue of a Glass Tube may be conveyed to any other bodies, so as to give them *the same* Property of attraction and repelling light Bodies, as the Tube does, when excited by rubbing'.[22] Because the electric vertue could be transmitted by materials that were not electrics, their nature no longer appeared to be relevant to the study of electricity. Moreover, because the materials that transmitted the vertue did not on their own attract or repel environing bodies, their nature was also irrelevant to the study of electricity. There was no other thing, apart from electrics and conductors, with essential properties that could have shed light on the experiments. Thus, the traditional philosophical search for the nature of things, and the discursive apparatus that was supposed to guide the philosopher in this search, seemed redundant in this context. Gray's letter to Sloane was written as an experimental report, albeit composed of somewhat surprising facts. It implicitly conveyed, however, claims about proper and improper manners of conducting philosophical research.

Attraction and repulsion were empirical facts, and so was the presence of the electric, including its attrition. Philosophical explanations were traditionally presented in the form of a learned commentary on these basic facts. Gray did not take part in this tradition. The authority of his explanation stemmed from the practical endeavor to reconfigure the familiar facts by way of experiment. His experimental work demonstrated in a technical idiom how the facts were interrelated. The experimental narrative conveyed these interrelations in an explanation that displaced traditional philosophical discourse. Common to his experiments was an

---

22 Gray 1731, pp. 18-9; italics added.

apparatus with a basic linear design. On one side he placed a glass tube, an electric body that had already been suggested in Descartes' *Principles.* He attached to the electric another body of linear shape, which formed the crux of the experimental setting, and added some very light bodies close to its other edge. His report then related how far the mediating body could be extended. He reported to have initially used a cork as a mediating body, but then replaced it with a short fishing rod, a 14 foot rod, and finally a 20 foot pole. The scale was gradually extended, 'so that it was not at all to be doubted, but with a longer Pole the Electricity would have been carried much farther'.[23]

Gray used, in addition, various materials – rods, sticks, poles, brass and iron wires, packthreads and silk cords – to examine their quality as electrical conductors, though most importantly 'Care was taken that the Tube was far enough off from having any Influence upon the Leaf-Brass [i.e. the attracted body], except what passed by the Line of Communication'.[24] These trials enabled him to simultaneously refine the experimental technique and his new explanation of electrical phenomena. For example, in one of his trials the communication of electricity was not followed by the expected result of attraction and repulsion. It happened when the 'line of communication' was suspended horizontally along a supporting beam. The experiment might have seemed to be merely an unsuccessful trial to 'communicate' the 'electrical vertue'; but instead, Gray argued that 'when the Electric Vertue came to the Loop that was suspended on the Beam, it went up the same to the Beam; so that none, or very little of it at least, came down to the Ball [of cork, which served as a touchstone for attraction]'.[25] As Gray noted in another report, one of the problems in reproducing electrical experiments was that the earth absorbed electricity. In virtue of the earth's magnitude, it was practically impossible to identify the trajectory of the 'vertue' and to detect the predicted effects.[26]

Philosophers before Gray noticed that the electrical phenomena depended to some extent on environmental conditions. Gilbert, for example, discussed the impact of humidity on electrical attraction, and considered it to be one of the factors that distinguished electrics from loadstone. Humidity weakened attraction, he explained, for various

23 Ibid., p. 22
24 Ibid., p. 27.
25 Ibid., p. 25.
26 Gray 1732a, 1736.

reasons: 'in thick weather light objects are harder to move,...the effluvia are stifled, and the surface of the rubbed body is affected by the vaporous air, and the effluvia are stopped at their very origin'. He concluded that electrics that 'do not so readily collect the humid air' were more effective than those 'which collect the condensed moist air on the surface'.[27] Gray's report from 1731 suggested a different explanation. Water conducted electricity, and for this reason the expected effects did not occur in a humid environment.

The relative ease with which materials could be tested for conductivity transformed prediction into a highly useful research strategy. Gilbert, or for that matter, his predecessors, could not test the properties of the effluents that supposedly characterized the nature of the electric body, though they clearly did not consider that to be a weakness. Gilbert proposed a reasonable explanation for the impact of humidity, which formed an integral part of his reasonable and highly innovative magnetic philosophy. Although Descartes did not address this specific phenomenon, he could convincingly argue, not unlike Gilbert, that it stemmed from the relatively high density of the third element that comprised the humid air, which effectively hindered the expansion of the second element that filled the porous electric. Their theories were designed to provide appropriate definitions of nature, rather than tools of prediction. Gray's report form 1731, by contrast, offered an explanation that encapsulated a tool for predicting electrical phenomena.

The difference between Gray's explanation and Gilbert's – or for that matter, Descartes' – did not pertain to formal definitions of causality. For example, their explanations aimed to identify the efficient or instrumental cause of the phenomena. Gilbert clearly sought to understand how humidity interfered with the special efficacy of the electric body. Descartes' explanation, as shown above, was primarily concerned with the mechanism that generated attraction and repulsion. Furthermore, the difference between their explanations did not lie in formal epistemic procedures. Gilbert, Descartes, and Gray drew inductive inferences from empirical evidence. They also deduced specific implications from general theoretical considerations. Induction enabled them to adjust theoretical abstractions to specific empirical conditions. Deduction enabled them to account for specific phenomena in a coherent and systematic manner.

---

27 Gilbert 1941, p. 91.

The novel characteristics of Gray's electrical research originated in a new conception of the aim of causal explanation. The aim was to understand how empirical properties were correlated. Electrical conductivity offered a solution to this problem, and thereby acquired three interrelated meanings. First, electrical conductivity denoted a fact. It was presented in Gray's reports as a physical process that either occurred, or failed to occur, under specific observable circumstances. Second, electrical conductivity denoted an experimental technique. The researcher who studied the facts of conductivity had to learn how to cause their occurrence. Third, electrical conductivity denoted a philosophical tool, namely a causal explanation of electrical attraction and repulsion.

The three interrelated meanings of electrical conductivity originated from various sources, such as experience, abstract reasoning, the design and performance of experiments, and philosophical literature. However, none of these resources sufficed to establish any one of the meanings. Although the glass tube, the conducting material, and the motion of attraction or repulsion could be easily observed, the process of conductivity itself transcended the bounds of empirical evidence. Inferences from the evidence involved abstract reasoning; yet logic did not explain how the pieces of evidence cohered. Technical dexterity was crucial for demonstrating electrical conductivity; though it did not on its own show how the demonstration could solve philosophical problems.

At its core, Gray's achievement consisted of transforming empirical, technical, and philosophical resources into a cultural product – his contribution to the philosophical study of electricity. In this respect, electrical conductivity was a human invention that was inherently embedded in the culture of natural philosophical research and in the broader context in which it was carried out. From the historical perspective on science as knowledge, Gray's contribution has seemed to be confined to the inadvertent discovery of some phenomena. As Cohen has noted, Gray was 'an empiricist whose ability lay in uncovering many new sorts of phenomena, [but he] did not make any notable contributions to electrical theory'.[28] From the perspective on science as causal explanation, it appears that Gray's discovery of new sorts of phenomena was the result of applying his novel approach to the philosophical problem of relating evidence with theory.

---

28 Cohen 1956, p. 296. See also Heilbron 1979, p.247; Home 1981, p. 42.

The Institution of Electrical Conductivity

The discovery of electrical conductivity was historically significant, as previously discussed, because it offered an alternative to the age old philosophical endeavor of explaining electrical phenomena by way of defining the nature of the electric body. Moreover, it enabled other researchers to expand upon Gray's original work. There were three times as many researchers publishing on electricity in the 1740s as there were in the entire 17th century. The new breed of 'electricians', as they were called by the late 1740s, included wandering lecturers, members of national academies and philosophical societies, as well as faculty members in colleges and universities. By the 1750s, books devoted to electrical research were regularly published by amateur researchers and academic lecturers. The research presented in these publications focused in one way or another on electrical conductivity. Experimentation with electrical conductors offered students an opportunity to develop careers in electricity.

In light of its three meanings, electrical conductivity set specific constraints on, and opportunities for, defining and solving new research problems. Researchers learned how to prepare and use conductors, refine their expectations of apparatus operation, and improve their explanation of both successful and failed experiments. Electrical conductivity thus became an integral part of routines that researchers followed in discerning the facts, explaining phenomena, designing experiments, and performing them. It became an integral part of the way to communicate these activities to colleagues, and explain them to the lay public. In this respect, electrical conductivity became a social institution. The transition from studies on the nature of electrics to studies of electrical conductivity marked an institutional change in the history of science.

The institution of electrical conductivity began to branch off in several overlapping directions. First, researchers sought to establish more rigorous and standard uses of electrical conductivity. In the early 1730s, Gray's colleague Granvil Wheler, Charles Dufay from the Académie Royale des Sciences, and the Royal Society's curator John Desaguliers, began to scrutinize distinctions between conductors, non-conductors, electrics, and non-electrics. To assess the efficacy of conductors, Dufay demonstrated the importance of insulating conductors with other materials, some of which had previously been identified as electrics. He then applied electrical conductivity to the systematic investigation of the dynamics of attraction and repulsion. Wheler conducted similar studies, and they both suggested that attraction preceded repulsion when the attracted body was not yet electrified. Repulsion occurred once it received the electrical vertue.[29]

Desaguliers applied the tools of rational mechanics with which he had earlier studied the flow of water in pipes to examine and model the flow of electricity through conducting material. He suggested, for example, that electrics, unlike conductors, transmitted electricity but could not receive it. He clarified the distinction between these materials in terms of 'saturation' and 'discharge', and suggested that conductivity involved the trade-off of electricity between saturated and non-saturated materials. He further explained that insulators were necessary for preventing the flow of electricity outside the experimental system through non-saturated materials in the environment.[30]

Other projects focused on electrical conductivity as an experimental technique and used it to expand the scope of electrical research. Experiments were conducted to explore a new means to amplify electrical effects by devising a more omnipotent system of electrical conduction. A celebrated device of the mid 1740s – the Leyden jar – consisted of a glass container filled with water. The glass was the electric body that received its electricity from a prime conductor, and the water, which was by then commonly known to be a good conductor, was used to discharge the saturated glass container. The sparks, blazes, and explosions that were generated by the electric jar defied common sensibilities.[31] A 1745 issue of *Gentleman's Magazine* reported that the effects were 'so surprising as to awaken the indolent curiosity of the public, the ladies and the people of quality, who never regard natural philosophy but when it works miracles. Electricity became all the subject in vogue, princes were willing to see this new fire, which a man produced from himself and which did not descend from heaven'.[32]

Another area of interest to electrical researchers and their new audience pertained to the vital powers of electricity. Gray inaugurated an inquiry to the electricity of life in the early 1730s, testing living bodies for their conductivity. As he reported to the Royal Society, visitors regularly attended experimental performances at his Charterhouse residence in

29 Dufay 1933, 1733-4; Wheler 1739-40

30 Desaguliers 1739-40, 1739-40a, 1763, ii, pp. 323-31.

31 Kuhn 1962, pp. 52-65; Heilbron 1979, pp. 309-18.

32 *Gentleman's Magazine*, Vol. 15, p. 194.

London to watch electrified schoolboys.[33] Hanged on insulating rods and attached to an electric body, the boys were turned into living conductors (Figure 2.2). The spectacle, which quickly diffused through the continent, generated a new imagery of the physical properties of psychic experiences.[34] Attending the show in Paris, one witness wondered how 'enthusiasm is, if I may so express myself, the image of the electric shock: it is a fire which is speedily communicated, which instantly embraces the spectators' imagination, which enflames their soul, which forces their hear to sensibility'.[35] Electrical conductivity additionally suggested therapeutic uses. In 1745, Cromwell Mortimer, physician and secretary of the Royal Society, professed that 'electricity may be used medically, in order to renew and regenerate a proper Quantity of vital Fire, such as is necessary for the conveniently carrying on, and performing the animal Functions'.[36]

The new science of electricity established itself at the crossroads of various fields of learning and polite culture. The old debates about the nature of electrics could not, and arguably did not seek to become so concretely popular. They displayed a form of reasoning that was too abstract to be infused into the experiences of everyday life. By virtue of its explanatory and experimental meanings, electrical conductivity conveyed the philosophical notion of ordered universe in a public affair that could be experienced everywhere. As Joseph Priestley pointed out in the first treatise ever published on the history of electricity,

> electricity is no local, or occasional agent in the theatre of the world. Late discoveries show that its presence and effects are everywhere...It is not, like magnetism, confined to one kind of bodies, but every thing we know is a conductor or non-conductor of electricity.[37]

---

[33] For Gray's electrical showmanship, see Schaffer 1983, 1997; Ben-Chaim 1990.

[34] Chipman 1958; Heilbron 1979, 247-8.

[35] Jean-Georges Noverre, *Letters sur la danse*, 1760. Quoted from Hobson 1982.

[36] Mortimer 1745, p. 479.

[37] Priestley 1775, i, p. xiv; McEvoy 1979.

2.2 Following Gray's shows in London, electrical conductivity excites a Parisian salon. Traditional philosophical discourse is represented by the textbook (at the center forefront), overshadowed by a new mode of experimental learning (Nollet 1746, by permission of the Houghton Library, Harvard University)

The Historical and Cultural Contexts of Gray's Discovery

Stephen Gray was the first person in the history of science whose entire research career evolved predominantly in relation to the study of electrical phenomena. Electrical research as a vocation did not exist before the 1730s. Few philosophers addressed the subject, and even those who did considered it to be peripheral to their careers. A science of electricity as an organized activity that spanned over a significant portion of careers of individual researchers came about soon thereafter, as previously discussed.

The new science rapidly distanced itself from the traditional philosophical discourse on the nature of electrics, though Gray himself could not possibly envision a career based simply on negating this tradition. A more complete historical picture of the institutional change that his discovery involved hinges therefore, on understanding the origins of his innovative research work, and especially the circumstances that formed the constraints and opportunities that directed his efforts.

Gray began his career as an amateur naturalist in the early 1690s, while managing his family dyeing business in Canterbury.[38] Following in the footsteps of his elder brother Matthias, he purchased some optical instruments and initiated studies in microscopy and astronomy. Several years later he established connections with John Flamsteed, the Astronomer Royal at the Greenwich Observatory. The latter's recommendation led to Gray's new appointment in the early 1700s as an assistant to Roger Cotes at the newly founded observatory at Trinity College, Cambridge. A student at Cambridge during these years, William Stukeley later recalled in his family memoirs that in 1704 Gray 'showed us many times his electrical operations with a long glass tube'.[39]

Gray's attraction to the electric body was engendered by several circumstances that coincided at the Royal Society of the early 1700s – a cultural center in which the tastes of polite society and the values of philosophical learning shed new light on experimental research. By the turn of the 18th century, amateur naturalists in England routinely established informal affiliations with centers of learning, and began to form provincial associations which, through personal connections, linked them to more established and prestigious centers. These networks invigorated polite

---

38 For a biographical perspective on Gray's career, see Cohen 1954; Clark and Murdin 1979.

39 Stukeley 1883, p. 378.

culture and rendered it more accessible to those who associated the cultivation of learning with their personal rise in the social hierarchy.[40] Expressing contemporary sensibilities, *The Guardian* proclaimed in 1713 that 'it is a sort of Impiety to have not Attention to the Course of Nature'.[41] *The Spectator* advised that 'a man of Polite Imagination is let into a great many Pleasures that the Vulgar are not capable of receiving...It gives him, indeed, a kind of Property in every thing he sees, and makes the most rude uncultivated Parts of Nature administer to his Pleasures'.[42] Derham's collected papers of the Royal Society suggested, in a quotation from Fontenelle's *Memoirs* of the Académie Royale des Sciences, that the experimental study of nature was morally inspiring because nature was not tainted by the 'effects of the Passion' and 'human Caprice'. 'The most curious Strokes of History should hardly be more curious than the Phosphorus...which Art discover'd by a near and diligent Scrutiny of Nature'.[43]

Gray established his initial connections with the Royal Society through a fellow townsman, Henry Hunt, an assistant to Robert Hooke and a minor official at the Society since 1676. The connection earned Gray his first copies of the *Philosophical Transactions* and his acquaintance with secretary Sloane. Gray's first experimental report on electricity, composed in 1708, reaffirmed the common affinity between experimental knowledge and the cultivation of polite character. The report was comprised of several related experiments which expressed his 'delight in the Knowledge of Nature', and were inspired by the 'Luciferous Experiments and Noble Discoveries' he had found in the Society's journal. The enchanting achievements that could be made in the 'Philosophical World', he added, 'Incouraged me to attempt the Persuance of those wonderful properties a little further'.[44]

Gray's special interest in electricity was aroused by new research projects that were initiated by Newton in his capacity as the president of the Royal Society since 1703. By then, Newton was widely acclaimed in England and abroad as one of the leading figures in the philosophical

---

[40] Allen 1976, esp. pp. 5-51; Plumb 1972, 1973; Porter 1980, 1981; Schaffer 1983; Stewart 1986, 1992.

[41] *The Guardian*, No. 169, September 24, 1713.

[42] *The Spectator*, June 21, 1712, p. 411.

[43] Derham 1726, i, The Preface.

[44] Chipman 1954, p. 46.

world. His election to the presidency of the Society coincided with his final preparations for the publication of his *Opticks*. The book was by and large devoted to a comprehensive and highly technical study of light and colors. Yet Newton decided to supplement this study with a series of 'Queries' which would present various aspects of his broader view of the physical universe, and would more directly link his optical research to his first major work, *Mathematical Principles of Natural Philosophy*, published in 1689. Following his election, he took advantage of every resource at his command to advance the reputation of his final major work.[45]

One of these resources was the service of Francis Hauksbee, a newly appointed curator of experiments at the Royal Society. Hauksbee was an acclaimed instrument-maker, and in his capacity as a curator he was responsible for preparing and operating experimental equipment which the fellows required for meetings at the Society, and for public lectures in other locations.[46] The collaboration between philosopher and technician focused partly on electricity. The ensuing dialogue between Newton's cosmological ideas and Hauksbee's experiments with electrics set the frame of reference in relation to which Gray pursued his new interest in the 'wonderful properties' of electrical phenomena.

Newton carried out a minor study on electrical phenomena in the mid 1670s.[47] He later drafted a 'Scheme' for the organization of research at the Royal Society, in which he associated the study of the gravitational force with the experimental study of electrics, magnetism, vapors, fumes, air and fire.[48] The 'scheme' was closely related to his broad vision of the physical universe as a system composed of material particles and forces. One force acted at a distance and worked to explain universal gravitation, as he demonstrated in *Principia*. Other forces, which he was unable to adequately identify, were associated with the more qualitative changes that characterized chemical and life processes.

In 'Queries' for his first edition of *Opticks* – the publication of which coincided with Hauksbee's appointment – Newton suggested that the process of illumination was generated by forces exerted over short

---

45 For detailed studies of Newton's career as the president of the Royal Society, see Westfall 1980, pp. 627-697; Heilbron 1983.

46 Heilbron 1979, pp. 229-34, 237-41; Westfall 1980, pp. 632-5; Stewart 1986.

47 Newton 1959, i, p. 365.

48 Brewster 1831, p. 103; Westfall 1980, pp. 632-3.

distances. The new curator undertook the task of investigating a particular case that appeared to involve these forces, namely the glowing effect of phosphors. Hauksbee designed a mechanical device for rotating a phosphoric globe to demonstrate that light and 'cold' fire were effects of dormant forces actuated by the motion of the minute particles that composed it.[49] He later demonstrated that various electrics, such as wool, amber, and glass, emitted light when rubbed in the evacuated container of an air-pump.[50]

Hauksbee's demonstrations vividly illuminated the received view of electric effluents. They concomitantly illustrated Newton's speculation that the generation of electricity was closely related to the generation of light, and that these processes were brought about by short range powers that were perhaps constitutive of the material universe as a whole. This soon became one of the themes Newton discussed in his revised drafts for the General Scholium of the new editions of *Principia*, and for the 'Quiries' in *Opticks*.[51] One revision suggested that 'electric bodies by friction emit a subtile exhaltation or spirit by which they perform their attractions'; that 'this spirit [was] of a very active nature & capable of emitting light by its agitations; and that 'all bodies abound with such a spirit & shine by the agitations of this spirit within them when sufficiently heated'.[52] In another draft, Newton attributed vegetative processes to the electric spirit, noting that 'the electric vertue is invigorated by friction so it may be by some other causes. And by being stronger in the particles of living substances then in others it may preserve them from corruption & act upon the nourishment'.[53] Another revision proposed that the 'subtle spirit' was not only confined to the pores of 'gross bodies', but 'expanded through all the heavens'.[54]

Hauksbee invested his technical skills in devising experimental representations of the abstract philosophical ideas which preoccupied Newton. He carefully designed his experiments to draw attention to the medium in which the effluents formed their potent configurations, and

---

49 Hauksbee 1705, pp. 2129, 2135.

50 Hauksbee 1705a.

51 Hall 1959; Koyré and Cohen 1960; Guerlac 1964; Hawes 1968, 1971; Heilbron 1979, pp. 239-41; Home 1982, 1985.

52 Quoted from Home 1982, p. 195.

53 Ibid., p. 199.

54 Ibid., p. 192.

developed experimental techniques for the public display of these configurations. He thus replaced the electric glass tube he originally used with a glass globe nine inches in diameter. He then rotated the globe on an axle that was whirled by a system of two pulley wheels, connected by a belt, and rubbed its surface with his hand. In an early report to the Royal Society, Newton had already professed that 'nature is a perpetuall circulatory worker'. The 'vast body of the Earth...may continually condense so much of this Spirit as to cause it from above to descend with great celerity for a supply'.[55] Hauksbee accordingly construed, and represented the movements of light bodies inside and outside the glass globe in the shape of radial forces. Using threads that were attached at one edge to a disk, he displayed a geometrical representation of their spatial configuration. It was 'a plain Instance of a Repulsive and Attractive force', he explained, generating centrifugal and centripetal motions 'so that in these small orbs of Matter, we have some little resemblances of the Grand Phenomena of the Universe' (Figure 2.3).[56]

Gray read Hauksbee's experimental reports in *Philosophical Transactions*, but distanced as he was from Newton's initiatives at the Royal Society, he was apparently unaware of the philosophical authority that these reports were designed to affirm. The amateur naturalist repeated some of Hauksbee's experiments that did not require special technical resources that were at the curator's command, but questioned the adequacy of his interpretations. In his first letter to the Society on electricity from 1708, Gray specifically addressed the received view that attributed attraction and repulsion to the electric effluvia, but considered it a 'hypothesis' which his experimental trials could 'confirme or overthrow'. He noted, for example, that the positioning of feathers and other light bodies affected their motion in relation to one another, and not only to the electric body. One experiment showed that a feather, repelled from an electric glass tube, was 'fixed upon' his hand. The evidence suggested that it was not necessarily the properties of the electric body that generated the motion of surrounding bodies. Gray thus proposed that 'all bodies Emitt soe they receive part of the Effluvia of all other bodies that Inviron them'.[57]

Gray's report was not published by the Society, however, and Gray himself did not press his point further. He was nevertheless determined to

55 Newton 1959, i, p. 366.
56 Hauksbee 1707, pp. 2372-7; 1919, pp. 54-8, 74-5, 80-1, 139-43.
57 Chipman 1954, pp. 35-6.

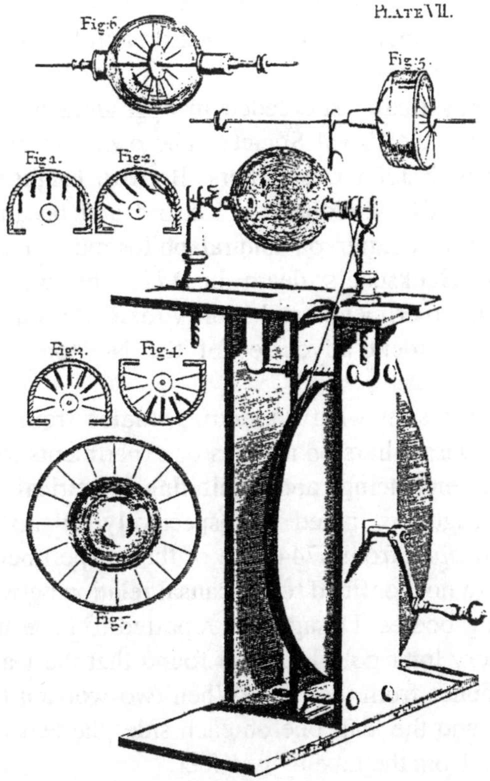

2.3 Hauksbee's electrical machine and his illustrations of the position of threads in reference to an electric glass globe. Spatial configurations of threads in reference to the electric body are supposed to represent the radial force of electrical attraction and repulsion (Hauksbee 1719, by permission of the Houghton Library, Harvard University)

continue his studies, and concomitantly sought to augment his connections with the Royal Society. In 1711, he asked Sloane to help him obtain admission into the Charterhouse in London, where he could 'make a further Progress...in some inquiries that might be of use'.[58] His appeal, however, was rejected. Another opportunity arose in 1713 when his friend Henry Hunt, a clerk at the Royal Society, died. The appointment to this position was negotiated on behalf of Gray by the new secretary, Brook Taylor, was

58 Cohen 1954, p. 42.

also denied. According to Taylor, Gray was a 'very fit person for the service of the Royal Society', but 'has such dreadful apprehensions of the presence of so many virtuosos'.[59]

In 1716 Gray finally succeeded in upgrading his intellectual and social status vis-à-vis the Royal Society. The occasion was brought by a new collaboration with John Desaguliers. Born to Huguenot immigrants, Desaguliers attended Christ College at Oxford and became a successful artisan and a favorite lecturer on natural philosophy in London's polite circles. Following Hauksbee's death in 1713, he was appointed the Society's curator of experiments, and shortly thereafter built his reputation as one of the most ardent advocates of the Newtonian philosophy in England.[60]

Resuming his somewhat dissenting studies from the mid 1700s, Gray worked with Desaguliers on a series of experiments to systematically test materials for enhancing and inhibiting electrical attraction and repulsion. As Desaguliers noted retrospectively in his *A Course of Experimental Philosophy* from 1744, one of their experiments showed that the phenomena were not confined to the causal relation between the electric and the surrounding bodies. Desaguliers reported to have used an electric glass tube and a very light gold leaf, and found that the leaf did not move when placed six inches from the tube. When two wooden boards were set alongside the tube and the leaf, one on each side, the leaf moved with the increase in distance from the tube by two feet.[61]

These studies led to Gray's first publication in the *Philosophical Transactions* in 1720. His account presented an experimental setting that was deliberately designed to question the age old conviction that electrical phenomena were caused by properties that distinguished the electric body from other materials. The trials he presented were not, however, set in the context of the traditional philosophical debate about the nature of the electric. This context had already been disregarded in Hauksbee's publications. Hauksbee assumed that the electric body was the principal cause of the phenomena; yet he did not purport to offer a discursive framework for the definition of its particular nature. His main task was to design experiments and produce representations that could vividly display a

59 Clark and Murdin, p. 388.

60 Stewart 1986, 1992.

61 Desaguliers 1744, i, pp. 18-9.

causal relation between the electric and surrounding bodies. Gray embraced the technical task, but refashioned its conceptual intention.

Thus, rather then focusing the experiment on a centered electric body, he reportedly arranged several light bodies in sequence, one after the other, and in various distances from an electric glass tube. The arrangement gradually formed an elongated representation of the transmission of the electrical 'vertue'. He introduced the term 'communication' to describe the formation of the path of transmission. He apparently did not coin this term. It was previously used by Hauksbee in a similar context, in his *Physico-Mechanical Experiments*. This work was published in 1709 by the Royal Society, presumably under the auspices of Newton, and was later incorporated into Desaguliers' first textbook.

During one of his experiments with threads that extended inside and outside of a rotating glass globe, Hauksbee reported to have observed that the external threads affected the position of the internal threads. He concluded that ''tis manifest that there's a Communication between the Medium without, and that within the Glass...Not only Communication, but a Continuity of the [subtle] Matter that occasions the Motion of the threads'.[62] Hauksbee was thus describing the changing motions of the threads in relation to one another. Gray too used the term to account for the changing spatial relations between non-electrics. Their accounts similarly attributed 'communication' to the attrition of the electric body. Their intentions were nevertheless considerably different, and the difference became more apparent as Gray further developed his experimental project.

The philosophical value of Gray's articles from 1720 was thus embodied in a novel experimental procedure that was designed to represent how the electric 'vertue' was transferred from one body to another. The next step consisted of replacing the discrete bodies that represented the path of electrical communication with longer lines of communication. His experiments eventually required a larger space than afforded in his residence at the Charterhouse, which he eventually received in 1716. Gray moved his apparatus to the estate of his friend John Godfrey, an amateur astronomer, and later to the estate of Granvil Wheler, a fellow of the Royal Society who soon became a close supporter of Gray's work. Back at the Charterhouse, Gray turned the new experimental technique into a form of public entertainment. He apparently invited Desaguliers to one of the shows in 1728. Desaguliers immediately announced before the Royal

62 Hauksbee 1719, pp. 141-2.

Society Gray's 'curious observations on electricity and the communication of electrical quality'.[63] A couple of years later, the novel spectacle was presented in a special conference of the fellows of the Society with a delegation headed by the Prince of Wales. Gray was then awarded the first Copley medal of the Royal Society. In another quarter of London's polite society, the novel spectacle received poetic acknowledgment – 'To break the sleep of elemental fire; To rouse the pow'rs that actuate Nature's frame'.[64]

---

63 Royal Society *Journal Book*, vol. 13, 1728, p. 330.

64 The epigrammatic praise of Gray's work was composed by Anna Williams, his assistant at the Charterhouse, but was apparently rephrased by Samuel Johnson. Clark and Murdin 1979, p. 404; Boswell 1964-71, i, pp. 19, 232, ii, p. 26.

Chapter Three

Explanation and Experimentation Transformed: Newton's Studies of Spectral Colors

Introduction

The previous chapter has shown that the discovery of electrical conductivity marked out a turning point in the history of electrical research in virtue of the new mode of causal explanation it comprised. Rather than addressing the nature of the electric body, the new mode involved explanatory and experimental practices that represented how the electric body, and the phenomena with which it was traditionally associated, were correlated.

These practices were inseparable form one another. Experimental techniques provided the means to generate representations that conveyed the causal explanation. The latter provided a theoretical account of experiments and their results, and explained why the techniques were successful, or alternatively, how they failed to generate predictable effects. The novel package of research practices was gradually institutionalized, as more and more researchers considered them as setting constraints on, and offering opportunities for the development of new research projects.

The ensuing break with the philosophical discourse on nature has not been adequately explained within the framework of the central dogma of science as knowledge. Gray's discovery did not refute old theories. Rather, it demonstrated how redundant and irrelevant they were. The transition from the traditional mode of explanation to the new one was the result of a cultural change. It could not have come about as merely a result of search for knowledge.

Gray presumably did not need to be told that the study of nature must be rational and based on experience. He needed, however, to be part of the cultural setting in which the new kind of explanations were publicly transacted to be able to take active part in their advancement. For this reason, it has been necessary to examine how Gray became familiar with the new mode of explanation. The final section of the previous chapter has shown that this setting evolved around Newton's initiatives at the Royal Society. It appears, therefore, that Newton's work was decisive in establishing the new mode of explanation.

The present chapter examines Newton's contribution to this process of cultural change by focusing on his early research on light and colors. Newton regarded these studies as one of his most important achievements. In the autumn of 1669, when Newton was appointed Lucasian Professor of natural philosophy at Cambridge, with barely five years of research experience behind him, he chose to deliver his inaugural series of lectures on refractive optics. His theory of light and colors was the subject of his first publication – an article in the *Philosophical Transactions* from 1672. During the early part of the 18th century, his optical works became associated most intimately with his reputation as England's leading advocate of experimental philosophy.

Newton's studies on mechanics and optics have often been portrayed as having brought the Scientific Revolution to its most mature and comprehensive, if not its final phase. His research on spectral colors in particular has been appraised as one of the most salient contributions of the 'century of genius' to the development of experimental research in physics. As Westfall has pointed out, 'no other investigation of the seventeenth century better reveals the power of experimental enquiry animated by a powerful imagination and controlled by rigorous logic'.[1] Newton's new theory has been accordingly construed by historians as one of the finest examples in early modern history of science as a quest for knowledge. Since the present book proposes an alternative to the dogma, received historical accounts of Newton's theory deserve close scrutiny.

The assumption that the scientific value of Newton's theory was principally derived from its empirical demonstration can be traced to the writings of Mach, one of the founders of modern positivism. In *The Principles of Physical Optics*, Mach professed to have extricated his exposition of the historical development of the science of optics from what he considered the 'metaphysical ballast' that had often dominated research before the rise of modern physics.[2]

---

1 Westfall 1980, p. 164.

2 Mach 1926, p. vii.

In accordance with his positivist perspective on science, Mach has pointed out that empirically confirmed principles were gradually formulated since antiquity on the basis of the assumption that the sunlight was homogeneous. Then came Newton, who showed that the length of the spectrum of colors following the refraction of sunlight through a prism was about five times wider than that of an unrefracted beam. This evidence suggested that the beam of sunlight was not uniformly refracted. It appeared that variation in refraction corresponded to variation in color.

The next step in Newton's demonstration, according to Mach, consisted of submitting the different parts of the spectrum to refraction by a second prism. The evidence then showed that the refraction of these parts varied in accordance with their color. The conclusion that logically followed from these experiments was that light was composed of unequally refrangible components that were distinguishable by color. This was Newton's principal theoretical claim in his first publication in 1672.[3]

Following Mach's reconstruction, studies of Newton's early notebooks have proposed detailed narratives of the development of Newton's innovative work.[4] On the basis of these studies, Shapiro more recently concluded that Newton 'discovered his theory in two distinct phases'. First, in his notebook on his studies at Cambridge since 1661, titled *Quaestiones quaedam Philosophicae*, he recorded his discovery of the unequal refraction chromatic rays (*c.* 1664-1665). The second stage began about a year later, when Newton conducted a series of experimental studies on the refraction of sunlight and the problem of chromatic aberration in refractive telescopes. In this context he elaborated the theory of the heterogeneous composition of light.[5]

Two notable features of the received account of Newton's theory highlighted its status as knowledge, and therefore, as a contribution to science as knowledge. First, the theory originated in empirical discovery – the unequal refraction of chromatic rays – that set his work beyond what was known by his predecessors. The second feature pertains to the problem of explanation. Once Newton made the empirical discovery, he did not

---

3 Ibid., pp. 84-6. This reconstruction is reaffirmed in Koyré 1965, pp. 41-3; Westfall 1962, 1962a; Bechler 1974; A. Shapiro 1980.

4 Hall 1947-9; Hall and Hall 1962; Westfall 1962.

5 A. Shapiro 1984, pp. 11-2. The dating of the second stage is further examined in Guerlac 1983.

allow his research to be obstructed by the 'metaphysical ballast' that had often preoccupied his predecessors. Rather than proposing a causal explanation for the phenomenon he had discovered, he set out to develop a theory that could be corroborated by the empirical evidence that was available to him.[6] According to Shapiro, Newton proposed a phenomenological definition of light rays, and refrained from speculating about their nature.[7] As McMullin more recently explained, 'if the light-ray is no more than a geometrical device to calculate the path of transmission from source to screen [upon which color images were projected], the claim that sunlight consists of 'rays differently refrangible' might reasonably be said to be deducible from the prism experiments'.[8]

As will be shown in what follows, however, neither of these two features is historically accurate. The next section examines some aspects of philosophical studies of colors conducted before Newton, and shows that that the correlation between color and index of refraction – allegedly the first component of Newton's discovery – could not be considered a novel finding of the 1660s. Several scholars reported to have found a correspondence between different spectral colors and angles of refraction long before Newton ever did. Arguably the most detailed and rigorous account of the unequal refraction of chromatic rays was presented by Descartes in his *Meteorology* in 1637. Historical evidence shows that Newton initially closely followed several of Descartes' optical studies, including his study on spectral colors.

Newton's debt to Descartes undermines the credibility of the received historical account of his new theory. It does not, however, lessen the historical significance of his new theory. It suggests that the meaning and value of the theory were determined in the context of other 17th century philosophical studies on light and colors, rather than merely by abstract criteria of knowledge.

With the rise of mechanical philosophy in the 16th and the 17th centuries, the study of colors became intimately related to a series of questions concerning the relationship between cognition and the material world. The problem of causality and the proper philosophical way to address it were inseparable from these questions. The assumption that Newton – or

---

6 For a critical appraisal of this view, see Cunningham 1988, 1991.

7 A. Shapiro 1980.

8 McMullin 1990, p. 69.

for that matter, any other scholar of the 17th century – could make a significant contribution to the study of light and colors without taking active part in resolving these questions is, therefore, highly implausible.

The second section examines how Newton's solution to the problem of causal explanation evolved during his experimental research on spectral colors. It shows that Newton participated in the philosophical discourse on the nature of light, up to a point. His work from the mid-1660s indeed show the beginning of a break with traditional discourse. However, he by no means renounced the problem of causal explanation. The break with the age old philosophical mode of explanation marked the beginning of his endeavor to develop an alternative mode of explanation. This endeavor transformed the function of experiment in natural philosophical research, along with the traditional approach to the problem of explanation.

Change and Continuity in the Discourse on the Nature of Light and Colors

In 1680, when Hooke delivered his inaugural lectures on optics at Gresham College, he professed outright in an oratorical manner, suitable, perhaps, to this special occasion, that light was homogeneous. It was indeed 'the first thing in the World to which the Almighty Creator gave his fiat, when he made the World, fiat Lux, Let there be Light'.[9] His statement, at that time, was highly polemical. By emphasizing that light was a definite thing which God created, and philosophers studied, he questioned the plausibility of Newton's claim that light was heterogeneously composed of numerous kinds of things, or, more precisely, rays.

Prior to 1672, Hooke's assertion presumably would have been passed as a truism. Light undoubtedly was created by God, with the rest of the universe, and there did not seem to be a reason to believe that God created many different kinds of light. Philosophers concomitantly assumed that their learned discussion addressed a definite subject, rather than practically innumerably different subjects. In virtue of its uniform nature, light gave rise to different empirical manifestations of illumination in different circumstances. One of these phenomena was the rainbow iris, which was accordingly considered to be a specific 'modification' of light.

9 Hooke 1969, p. 72.

With other illuminations in the sublunar sky, such as auroras, moon rings and halos, the iris was traditionally classified as a meteorological phenomenon.

These general philosophical considerations structured the search for the causal explanation of spectral colors. Following Aristotle, late medieval scholars were primarily concerned with the nature of light. Those who set out in addition to explain the formation of rainbows, applied their definition of the nature of light to examine this particular modification, and arrived by this route to a definition of the nature of rainbow colors. Within this discursive framework, philosophers commonly distinguished between the illuminated colors of physical bodies on the one hand, and colors of the rainbow on the other. The former's colors were considered real properties, in contradistinction to the latter's, which were apparent. The rainbow colors, they taught, were generated by various degrees of mixtures of transparent air and relative darkness of clouds. The outer band of the rainbow was bright red because it was the largest, and comprised a mixture that was dominated by the transparent air. The inner band was dark blue because it comprised a mixture that was dominated by the relatively less transparent media of the condensed clouds.

During the 14th century, experimental devices such as transparent crystalline spheres, prisms, and water containers were introduced to simulate the formation of a rainbow. These experimental simulations enabled scholars to achieve more accurate representations of the order of bows and their geometrical formation. The philosophical interest in the simulated spectrum remained minor, however, especially in comparison to the growing interest in the relationship between the practical and philosophical aspects of vision, its artificial aids, and the techniques of visual imaging, which were engendered by the Renaissance of arts and the rise of humanism in the 14th century.[10]

With the development of mechanical philosophy in the 17th century, the age old distinction between real and apparent colors was rejected, most notably by Descartes. As a result, peripatetic explanations of apparent colors in the rainbow were abrogated. Critics of the scholastic tradition nevertheless tended to preserve the hierarchical structure of the discourse on particular natures in which general considerations set the precedent for more concrete considerations. Thus for example, Descartes'

---

[10] Boyer 1987, 47-9; Lindberg 1978, pp. 355-62.

arguments against received views on rainbow colors did not originate in his meteorological studies. Instead, they contributed to his philosophical discussion of a much more general and fundamental issue, namely the interaction between the immaterial soul and the material world. As he explained in his *Treatise on Man* and *Dioptrics*, material objects in the natural world affected the sense organs and thereby triggered the formation of mental images, which were then processed by intellectual faculties. It had to be admitted that visual sensations could not be considered properties of physical objects. Since light was a physical object, its nature was material. He could thus persuasively suggest that the sense of spatial orientation of a blind person did not essentially differ from visual orientation, since in both cases perception decoded the impact of material entities on sensory organs.

In virtue of the affinity of the subject to human physiology and knowledge, as well as to the analysis of matter and its motion, Descartes addressed the nature of light in various contexts. In his *Principles of Philosophy*, he defined the nature of light in reference to his theory of the grades, or elements, of matter, as has been noted in the previous chapter. Light was formed by matter endowed with 'so much force of agitation that, by colliding with other bodies, it is divided into particles of infinite smallness'.[11] In his *Dioptrics*, by contrast, he sought a definition that could be treated with the rigorous tools of geometry and rational mechanics. This was an important consideration for Descartes, since he was then affiliated with a circle of scholars, including Isaac Beeckman, and later, Marin Mersenne and Claude Mydroge, who were actively interested in the application of geometry and mechanics to the study of vision and visual aids.[12]

In an attempt to establish a unified framework for philosophy, geometry, and mechanics of light, Descartes introduced analogies and idealizations that provided the framework for his explication of the nature of light. He thus acknowledged that his *Dioptrics* was not intended to explicate 'what is its true nature'. His philosophical discussion aimed, in this specific context, to 'help to conceive of it [i.e., the nature of light] in a manner that seems to me most convenient, in order to explain all those of its properties that are known to us from experience, and then to deduce all the others that cannot be noticed so easily'. Descartes modeled, in part, his

---

11 Descartes 1983, p. 110.

12 Ribe 1997.

optical research on the most prominent discipline of the mixed mathematical sciences, namely astronomy. He thus noted that 'in this I imitate the astronomers, whose assumptions are almost all false or uncertain, but who nevertheless draw many true and certain consequences from them because they are related to various observations they have made'.[13]

Despite their fictitious character, Descartes' mechanistic explications of the reflection and refraction of light were widely recognized as important achievements for mechanical philosophy in its struggle for supremacy.[14] Ironically, these explications led to his demonstration of unequal refraction of spectral colors in his *Meteorology* – the phenomenon which, as noted earlier, has been construed by modern scholars to have provided Newton with the empirical foundation for his theory.

Descartes was by no means the only scholar who noted the unequal refraction of light in simulations of the rainbow iris before Newton. Marcus Marci, a professor at the University of Prague, presented this phenomenon in his *Thaumantias, Liber de arcu coelesti* in 1648. In this work, Marci reported on a series of experiments with prisms showing that each color of the spectrum was directly related to a definite index of refraction. As he stated in his concluding Theorem XVIII, 'it is not possible either to have the same color with a different refraction, or different colors with the same refraction'. He also found that a second refraction of monochromatic rays did not produce any alterations in colors.[15]

English mathematician and astronomer Thomas Harriot reported to have obtained measurements of prismatic refraction that suggested a correspondence with color and refracting media. Harriot obtained measurements of refraction through different media by using a hollow glass prism filled with different transparent fluids. In his mathematical treatise on light, the Jesuit professor at Bologna, Francesco Grimaldi, focused on the dispersion of spectral colors. He demonstrated that the colors did not change by further refractions.[16]

Descartes went further than these scholars and offered a philosophical explanation of unequal refraction. His investigation of spectral

---

13 Descartes 1985, i, p. 152-3.

14 For historical reconstructions of Descartes' derivation of the law of refraction, see Buchdahl 1969, Ch. 3; Schuster 1977, Ch. 4; Sabra 1981, Chs.1 and 4; Costabel 1982, pp. 63-76; Eastwood 1984; Shea 1991, Ch. 10.

15 Marci 1968. See also Ronchi 1970, pp. 156-8; Aiton 1970; Lohne 1968.

16 Lohne 1959.

colors as whole was of special importance, as he wrote to Mersenne in 1629, 'for I have decided to exhibit it publicly as a sample of my Philosophy, and to hide behind the canvas to listen to what people will say about'. Several years later he noted that his account of the rainbow comprised an exemplary case of the philosophical method he devised.[17]

Descartes displayed his methodical approach to the problem by presenting at first a series of observations and measurements of the order of the rainbow colors. These suggested that the rays that reached the eye with sufficient intensity subtended the angles of approximately 42° and 52°, respectively, with the continuation of the line joining the sun to the observer. At these angles of minimum deviation, the color-image was always red (At segments D and K of the primary and secondary rainbows, the angles marked DEM and KEM, respectively, in Figure 3.1 on next page). The color-images disappeared when the angles were slightly larger, and they changed from red to yellow and then blue, as the angles were gradually decreased.[18]

In the second stage of his demonstration, Descartes obtained more controlled evidence by refracting a beam of sunlight through a prism (MNP, in Figure 3.2 on p. 53). When 'the rays of the Sun ABC cross MN at right angles', he noted, 'they do not undergo any noticeable refraction there', but then 'suffer a reasonably large refraction on leaving through NP.' He then covered the prism and allowed the light to pass only through 'a narrow opening DE', and observed 'that the rays, passing through this opening and from there making for the cloth or paper FGH, paint all the colors of the rainbow on it', red always appearing 'at F, and blue or violet at H'.[19] The experimental results suggested, then, that chromatic rays were not parallel to one another and that the angle between the red and blue rays was the largest.

Further experiments demonstrated that these results did not depend on the size or material composition of the refracting media. The same results were also obtained with a glass prism and a flask of water. In accordance with his measurements of primary and secondary rainbows, Descartes further demonstrated how the refracting prism should be set to ensure that the position which the refracted rays reached indicated minimum deviation. These measurements undermined the plausibility of peripatetic accounts, he

---

17 Garber 2001, pp. 39-44, 95-101.

18 Descartes 1998, pp. 85-6.

19 Descartes 1998, pp. 87-8.

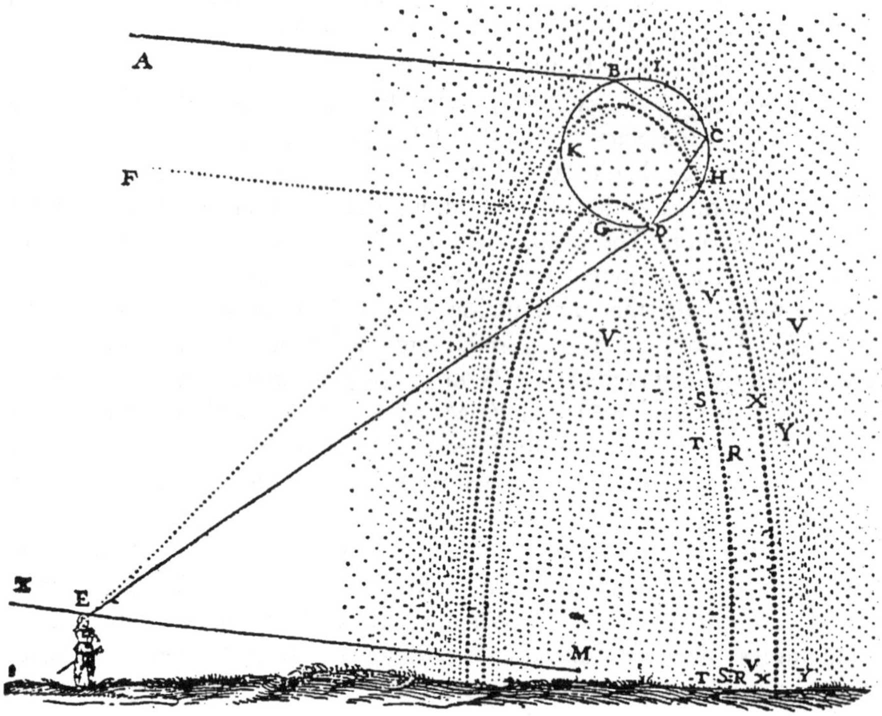

3.1 Descartes' geometrical illustration of the refraction of sunlight through clouds, followed by primary and secondary rainbows (Descartes 1677, by permission of the Houghton Library, Harvard University)

explained, because light and shadow surrounded the red and blue poles of the spectrum 'in the same way'. His experiments demonstrated in addition that a single refraction, under these conditions, was a necessary and sufficient condition for the appearance of colors.

In the next stage of his demonstration, Descartes appealed to 'the nature of light', 'as I described it in *Dioptrics*, namely as the action of motion of a certain very subtle matter'. An analogy with rolling balls suggested a convenient idealization (Figure 3.3 on p. 54). The forward edge of the incident beam of sunlight could accordingly be envisaged as a series of balls adjacent to one another. Descartes stipulated that the balls of the incident beam traversed with the same velocity. As they obliquely passed

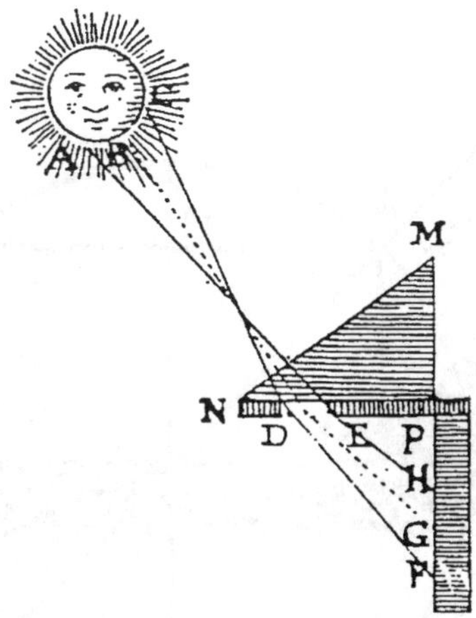

3.2 Descartes' diagram of the refraction of sunlight through a glass prism (Descartes 1677, by permission of the Houghton Library, Harvard University)

through the transparent medium of the glass prism, however, they acquired a rotary motion, since the movement of the side of the ball that 'encounters it [i.e., the prism] first, is retarded', while that of the other side 'still continues'. The balls rotated in the same direction, and transmitted their motion to adjacent balls. As a result, the balls at the one side of the refracted beam gained greater rotary motion than those at the other side. The greater their motion was, the swifter their pass through the prism became. Their angle of refraction, accordingly, was relatively small. Slower balls, by contrast, suffered greater refraction. This in turn explained how the order of the projected colors corresponded to variation in the magnitude of refraction. Descartes' mechanical account further explained how different media
changed the scale of the magnitude of refraction for the various colors but did not change their relative order or index of refraction.[20]

Descartes' statements on the nature of light and colors were by no means the last words on the subject. Practically every scholar of the 17th century who associated his career with mechanical philosophy found reasons to disagree with Descartes.[21] Moreover, the abolition of the peripatetic

20 Descartes 1998, pp. 89-90.

21 For studies of these debates, see Sabra 1967; A. Shapiro 1973.

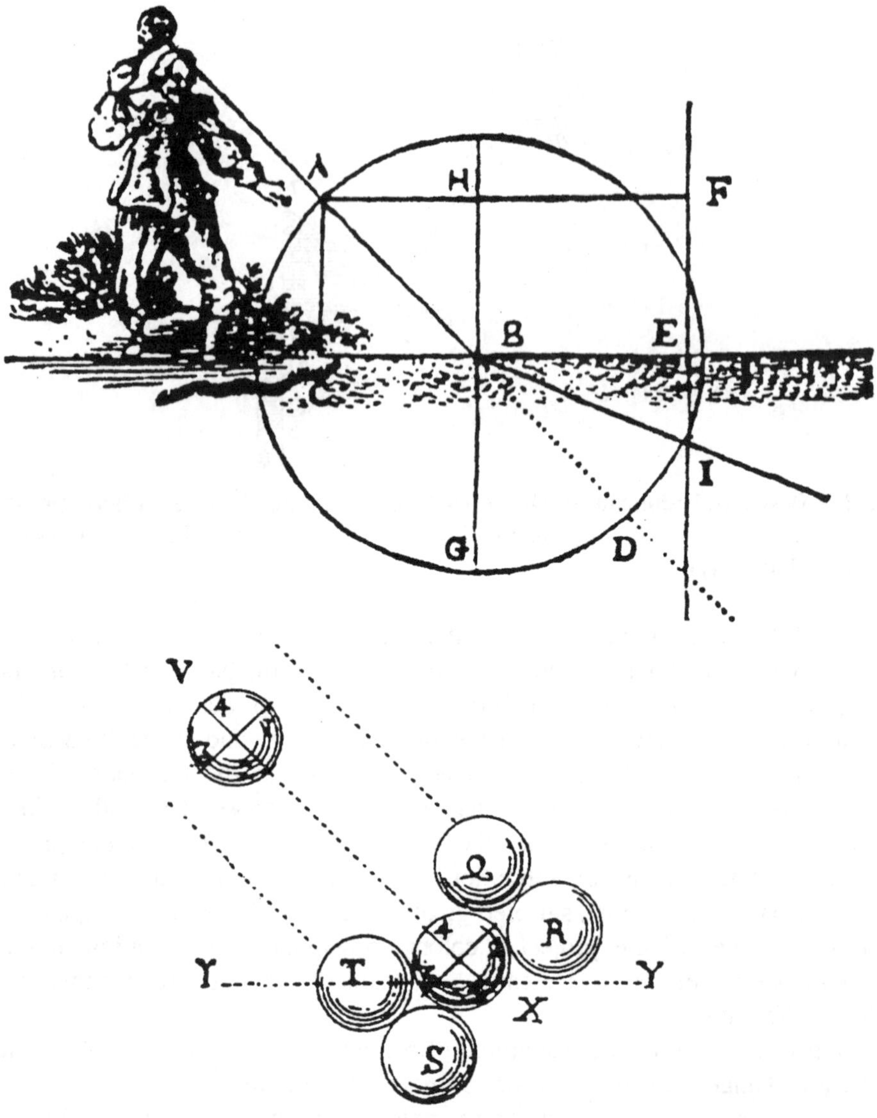

3.3 Descartes' illustrations of the trajectory of a tennis-ball as an analogy of the refraction of light (top), and of the mechanical model of an incident beam of light upon refraction (bottom) (Descartes 1677, by permission of the Houghton Library, Harvard University)

dichotomy between real and apparent colors formed a starting point for research projects that undermined the traditional assumption – which Descartes accepted – that the explanation of colors must be predicated on a definition of the nature of light. Since all colors were generated by the interaction between light and matter, the study of colors could be considered the setting for new methodologies for the study of light and matter, rather than the other way around. In particular, the study of colored surfaces and liquids offered a new perspective that combined the traditional tools of the geometry of illumination with the tools of chemistry.

One research project on the chemistry of light was pursued by Thomas Willis, Sedleian Professor at Oxford, and was presented in his *Practice of Physick.* Originally published in Latin in 1659, this book was considered by English scholars to be a major contribution to experimental studies of medically related problems in chemistry and physiology.[22] The first part of this book, titled 'On Fermentation', included a study on spectral colors and colored liquids. In an inspection of the changing color of liquids, Willis noted an instance of unequal refraction of chromatic rays. He drew the attention of readers to the interaction of light with particles composing the liquid, noting that the latter, by transmitting 'variously the Rays of Light, many ways break or reflect them, and so make diverse appearances of colors'. The order of colors corresponded to the changing magnitude of refraction: 'when the Rays of Light pass through almost in right Lines, they make a clear color, like Spring-water'; when 'they be a little broken, the [originally transparent] Liquor grows yellowish; but being more refracted, they cause a red color'. The appearance of other colors, he suggested, could be examined 'after this fashion'.[23]

Willis' experimental report on the unequal refraction of chromatic rays formed only a minor part of his project. His main interest pertained to the properties of blood and other humors associated with the 'flame of life'. Robert Boyle's experimental research on colors formed an important link between the broad perspective on colors that was presented in Willis's *Practice of Physick* and the prism experiments exemplified in Descartes' *Meteorology.* A colleague of Willis at Oxford, Boyle took part in developing the study of colors on the assumption that processes of illumination were

---

22 Willis' research and reputation, especially in association with the Oxford Club of experimental philosophers, are detailed in Frank 1980.

23 Willis 1684, p. 41.

constitutive to the formation of life and were involved in the changes of material properties of liquids and solids. As he noted in his *Experiments and Considerations Touching Colors*, 'the Importance of examining the Colors of bodies' directly pertained to the investigation of 'the dispositions of parts of Bodies', a subject of special interest to 'Chymists'.[24]

Boyle's *Experiments and Considerations*, published in 1664, offered the most detailed study of colored surfaces and liquids to date. A considerable part of this treatise was devoted to a methodological discussion that addressed the definition of the nature of light and its function in the experimental study of colors. Boyle explicitly adhered to the Cartesian dictum that colors were sensory images rather than genuine properties of material objects. The 'new science' of colors, he professed, began with the abrogation of the peripatetic notion that 'Colors might in some sense be consider'd as a Quality residing in the body'.[25]

The progress of the 'new science' did not, however, hinge on replacing old with new definitions of the nature of light and colors. Quite to the contrary – Boyle pointed out – the search for new definitions hindered progress. It merely introduced unnecessary divisions between different 'sects', each of which was committed to its own definition. Some scholars followed ancient 'Atomists', and taught that color was 'a Corporeal *Effluvium* issuing out of the Color'd Body'. Others endorsed the position of '*Chymists*', who attributed 'the Origine of Colors to the Sulphureous Principle in Bodies'. Another favorite dictum was 'the Hypostatical Principle, *Mercury*'. There were, additionally, 'the *Cartesians*', with their '*Globuli*, which 'tis not so easy to evince'.[26] The 'Embracers' of these theories, he professed, 'by undertaking each Sect of them to explicate Colors indefinitely, by the particular Hypotheses they maintain, seem to hold it forth as the only Needful Theory about the subject, yet for my part I doubt whether any one of all these Hypotheses have a right to be admitted Exclusively to all others'.[27] These theories were hypothetical not simply because they could not be adequately proven; rather, they claimed to provide a definition of the nature of light or colors, when it was commonly

---

24 Boyle 1664, p. 4. For Newton's debt to Boyle's study of colors, see Hall 1947-9; McGuire and Tamny 1983, pp. 241-74.

25 Boyle 1664, p. 21.

26 Ibid., pp. 87-8.

27 Ibid., pp. 86-8.

acknowledged that human understanding cannot reach certain knowledge of particular natures.[28]

The notion that different colors were generated by different particular natures made sense in light of the peripatetic assumption that the color of a physical body was a real property of that body. The 'new science' discarded this assumption, and should have therefore discarded the notion of a correspondence between different colors and different natures. The sensation of color was always the product of a plethora of intertwining factors. The value of experimental research was not confined to the evidence it could yield in support of a given hypothesis. As Boyle endeavored to show, experiments were crucial to any inquiry that set out to examine how complex and varied the configuration of these factors were. He noted, for example, that the exertion of physical pressure on the eye could generate color-images in the dark. The appearance of colored objects in dreams similarly suggested that the 'Optick Nerve' could operate independently of external stimuli. These sensations were not generated by exposure to light, and therefore could not be attributed to the nature of light.

Numerous other observations suggested that the color of physical bodies regularly changed somewhat independently of the degree of their exposure to light. The color of the skin of animals and plants was obviously affected by physiological processes. Exposure to heat changed the color of many solid materials, and liquids changed color when mixed with one another. Spectral colors were traditionally observed in the rainbow, but apparent colors could also be observed in the flames of fire.[29]

Moreover, explanations of spectral colors that were predicated on a specific definition of the nature of light but that did not take into consideration the texture of refractive media were similarly highly misleading. As microscopic observations suggested, the surface of physical bodies was neither homogeneous nor completely stable. It consisted of 'a multitude of singly insensible Corpuscles' that formed highly complex configurations that could easily be transformed, even by their interaction with light. The factors that determined these configurations 'concur to vary the forms of Asperity that Colors so much depend on'. Addressing his critique to the Cartesians, Boyle cautiously warned that the 'Celerity or Slowness' of 'Globuli', and 'their Revolution or Rotation in reference to

28 Dear 1998, p. 166.

29 Boyle 1664, pp. 3-25, 82-92.

their Progressive Motion', could not adequately represent the interaction between light and matter.[30]

Boyle sought to dissociate his work from the tradition of philosophical explications of the nature of light and colors. Addressing his readers, he requested that his book be considered 'not as a Discourse written Principally to maintain any of the fore-mention'd Theories, Exclusively to all others, or substitute a New one of my Own, but as the beginning of a History of Colors, upon which, when you and your Ingenious friends shall have enriched it, a solid Theory may be safely built'.[31]

Boyle's 'History' included a long series of experimental narratives that concomitantly offered a large repertoire of simple experimental devices and apparatuses that students could employ to explore the conditions in which colors were formed. He recommended 'the Glass-prism to be the usefullest Instrument Men have yet imploy'd about the Contemplation of Colors',[32] but, unlike Descartes, he presented a series of experiments with several prisms, mirrors and focal lenses, colored glasses and other semi-transparent materials, which were applied to the comparative study of different images of the spectrum. Detailed reports addressed the changing distances between the source of light, the position of refractive and reflective devices, and the projected images.[33] This rich repertoire of experimental devices and practices formed an important part of Boyle's contribution to the 'new science' of colors. It set him apart from contemporary scholars who conceived of their authorship within the bounds of the discourse on the nature of things. However, experimental works were not simply intended to provide 'satisfaction of Fancy', Boyle warned. Their value was in 'their use, not their strangeness, or prettiness'.[34] The new authority of science clearly belonged to experimental explanations.

Newton's Experimental Explanation of Colors

Newton began his studies on light, vision, and colors in the mid 1660s, his fourth year as a student of Trinity College, Cambridge. He registered his

30 Ibid., pp. 26-9.
31 Ibid., pp. 89-90.
32 Ibid., p. 227.
33 Ibid., pp. 185-197.
34 Ibid., 'The Preface'; Golinski 1989.

studies in a notebook, titled *Quaestiones quaedam philosophicae*. The table of contents, which he arranged alphabetically, displayed the variety of natural philosophical topics he addressed at the time. His entry on colors, for example, was preceded by entries on comets and cold, and was followed by notes on corruption, creation, and condensation. His entry on light was preceded by an entry on levity, and followed by entries on matter, motion, memory, meteors, and minerals. His notebook thus formed a topically arranged commonplace book in which, for the most part, he recorded notes and comments pertaining to his readings.[35]

Newton's philosophical notes reflected the authors he had read. Upon reading Descartes, for example, his entry on vision included an analysis of the mechanics of illumination which focused on the motion of a single 'globule of light' in the ethereal medium.[36] The longest and most detailed entry on colors included a report of experiments suggested by Boyle. One trial consisted of pressing a finger on his eye, and was followed by a detailed record of the 'vivid impressions' of colors that resulted despite the absence of an external visual object. He would later express his debt to Boyle's work in reference to this trial in a letter to Locke from 1691, the year Boyle died.[37] Other experiments pertained to colored surfaces, rather than to spectral colors.

Several paragraphs in the entry on colors comprised his first record on the unequal refraction of chromatic light rays. As noted in the introduction of this chapter, this record has been identified by historians as Newton's first major optical discovery, the initial stage in the development of his new theory. This record was not, however, distinguished as being more valuable to Newton in the context of his notebook.

In the broader context of 17th century philosophical studies of colors, Newton's first record of unequal refraction was by no means original. As shown in the previous section, several authors had reported a correspondence between colors and indexes of refraction long before he began his studies. Newton might not have consulted all of these reports, but it is unlikely that he was unaware of Willis' *Practice of Physick*.[38]

---

35 The entire notebook is included in McGuire and Tamny 1983, pp. 330-465.

36 Ibid., p. 384.

37 Newton's letter to Locke, June 30, 1691, in Newton 1957, iii, pp. 152-154.

38 Newton apparently did not own a copy of this book. However, the Latin edition from 1659, titled *Diatribae duae medico-philosophicae*, was included, with two other books by Willis, in Isaac Barrow's library in Cambridge. Newton

Newton's debt to Descartes, however, was the most direct and formative, as a closer reading of his notes would indicate.

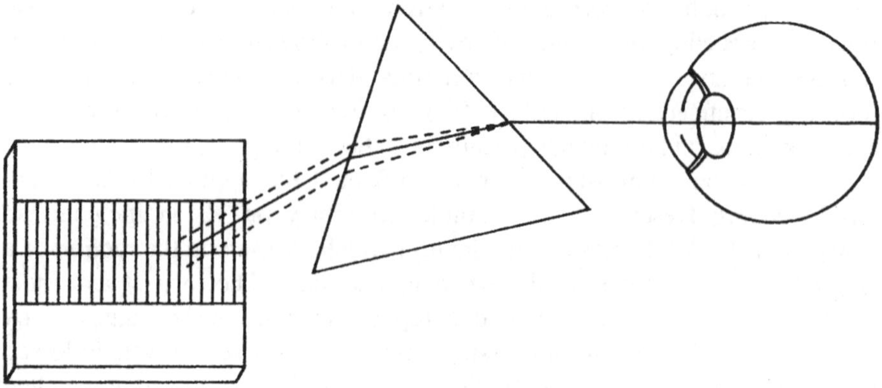

3.4 A reconstruction of Newton's illustration of a colored thread inspected through a glass prism (based on McGuire and Tamny 1983, pp. 434-5)

Newton's report on the unequal refraction of chromatic rays followed an experiment in which he observed a colored thread through a prism, half of which was blue, and the other half red (Figure 3.4). Newton might have conceived of this experiment after Boyle's extensive studies on colored surfaces, but his philosophical analysis and interpretation of its results were not foreshadowed by Boyle's accounts in his treatise on colors. Rather, they closely followed Descartes' methodical presentation in *Meteorology*. Newton thus began with a plain description of the trial followed by a geometrical analysis of its results. The color-images were displaced by the prism, and, since the angles of incidence of the reflections from the red part and the blue part of the thread were about the same, Newton concluded that 'ye rays wch make blew are refracted more y^{n} ye rays wch make red appears from this experimnt. If one hafe of ye thred *abc* be

definitely knew of this book by 1677, when, upon Barrow's unexpected death in May that year, he supervised the preparation of the catalogue of Barrow's library. But he probably had read this copy earlier, since he had extensively used Barrow's library until Barrow's death. For Newton's access to this library and its inventory, see Feingold 1990.

blew & ye other red & a shade or black body be put behind it then looking on ye thred through a prism one halfe of ye thred shall appeare higher y^{n} ye other. & not both in direct line, by reason of unequall refraction in ye differing colors'.[39]

The geometrical reasoning was followed by a philosophical interpretation in which Newton noted that 'slow' rays were refracted more than 'swift' ones. Swifter rays were less hindered than slower rays by the refracting prism, which acted as an elastic buffer – just as Descartes had detailed in his rolling balls analogy in his *Dioptrics* and *Meteorology*. Variation in speed caused variation in the magnitude of refraction and color.[40] Newton accordingly explained that the propagation of light through the air and its refraction by the denser medium of the glass prism affected the motion of the globules.[41] His brief account disclosed a highly informed and precise reiteration of Descartes' entire exposition.

The entries in his early notebook were followed by more extensive studies which Newton summarized in a separate treatise titled *Of colours* in 1666.[42] The treatise included many more experimental reports than the earlier notebook. These were arranged in numbered series, each of which related a small experimental project. Unlike his previous studies, this treatise primarily addressed spectral colors, indicating Newton's decision to examine the problem of colors in a context in which it was traditionally discussed by philosophers.

It could be expected, therefore, that in this new treatise he would continue to develop the ideas he found in Descartes' work, which addressed spectral colors rather colored surfaces. Newton, however, neither referred to his earlier notes on the mechanics of light particles, nor considered other ideas with which contemporary philosophers explicated the nature of light and colors in this treatise. The new treatise, which would eventually form the core of Newton's Cambridge lectures on optics, was written as if he did not know of Descartes' philosophy, or for that matter, of any other work that proposed a mechanistic explanation of optical phenomena.

The transition from the early notebook to the later treatise thus seems to have marked out a turning point in Newton's optical studies. From

---

39 McGuire and Tamny 1983, pp. 434-5.

40 Ibid.

41 Ibid., pp. 436-7.

42 Ibid., pp. 466-489.

the perspective on science as knowledge, the received historical view, as outlined above, suggested that at this stage, Newton progressed beyond his early empirical discovery and began to develop his theory. The path from an empirical discovery to a new theory has been construed by historians as an emblem of the method of scientific knowledge. Historical evidence shows, however, that Newton did not register an original discovery in his early notebook. Rather, he noted a phenomenon that had been recorded by other philosophers, and then explained it in a manner that revealed his debt to Descartes.

From the perspective on science as causal explanation, the transition appears to have been the outcome of Newton's strategic decision to change his approach to the philosophical study of colors. More particularly, the decision appears to have directly pertained to the relation between explanation and experimentation. In his previous notebook, he adhered to the Cartesian philosophy, whereas his later treatise was predominantly comprised of reports of experimental work.

Newton's strategic decision to change his approach to the philosophical study of colors raises two historical questions. The first pertains to his changing attitude towards the discourse on the nature of light. Why did Newton abandon his interest in the philosophical ideas he had earlier found in Descartes' writings? The second question addresses his research work. Once he separated Descartes' empirical exposition of unequal refraction from its philosophical interpretation, how did he utilize the empirical part during the forthcoming stage of his studies? Both of these questions address the intellectual and social aspects of Newton's career, and more specifically, his changing position vis-à-vis the public transaction of causal explanations of optical phenomena.

The answer to the first question must depend on indirect evidence. His decision to abandon Descartes' path may have been engendered by his reading of Hooke's *Micrographia*, published in 1665. Hooke rejected Descartes' rolling ball analogy, and adopted a wave analogy previously proposed by Hobbes.[43] Hooke thus proposed that light propagated 'just after the manner (though indefinitely swifter) as the waves or rings on the surface of the wave do swell into bigger and bigger circles about a point of it, where…the motion was begun'.[44] The spectrum, according to Hooke,

---

43 For a study of wave theories of light in the 17th century, see A. Shapiro 1973.

44 Hooke 1961, p. 57.

was generated by an 'orbicular' front, or 'pulse', traversing obliquely through the denser medium; the unequal advancement of the wave being the cause of the different colors.

Hooke's explanation thus offered an alternative to Descartes' mechanistic interpretation of the varying velocities of chromatic rays. Newton was apparently troubled by the ease with which different mechanistic analogies were adapted to explain the phenomena. In his copy of Hooke's *Micrographia*, Newton noted that the author's choice of the wave analogy led to the selection of empirical evidence that was consistent with it.[45] By 1666, Newton had already been concerned with the problem of the relation between explanation and experimentation. The explanation he proposed in his *Quaestiones* was consistent with the evidence he obtained; yet so was Hooke's. Moreover, in the absence of a tangible measure of the speed of light, or of the correlation between the properties of the prism and the deceleration of refracted rays, both explanations belonged to the class of sectarian hypotheses which Boyle had insisted on separating from the experimental study of colors. Experimental research seemed to be somewhat redundant if the ultimate aim of the philosophical study of nature was to propose an explanation that was merely consistent with the phenomena.

Newton's appreciation of Isaac Barrow's work may have further entrenched his decision to abandon explanations that rested on a definition of the nature of light. Barrow was the first professor to receive the Lucasian chair of mathematics at Cambridge. He delivered his inaugural series of lectures in the beginning of 1664. Newton apparently attended some of the lectures Barrow later delivered on the geometry of illumination and vision at Trinity College. Barrow was impressed with Newton's talent and interest, and when Newton became a Major Fellow of Trinity in 1668, he assisted Barrow in revising optical lectures which were brought to publication in late 1669.[46]

Barrow's lectures were devoted almost exclusively to geometrical studies on the formation of visual images. He taught the laws of reflection and refraction after Kepler and Descartes. He proposed a mechanical theory of light based on the works of Descartes and Hobbes, as well as on effluvial notions of emission. The professor stressed, however, the importance of demarcating the geometrical analysis of illumination and vision from the

---

45 Hall and Hall, 1962, p. 412.

46 Westfall 1980, pp. 99, 202-4; A. Shapiro 1990, pp. 111-3.

explication of the nature of light. The latter consisted at best of 'plausible conjectures instead of certain principles', he noted in his conclusion.[47]

Barrow accordingly understated the value of the philosophical study of colors. Predicated on the philosophical definition of the nature of light, the explanation of rainbow colors was farther remote from his principal interests as a mathematician. Barrow offered some 'guesses' on this subject 'by way of digression' from the geometrical analysis of the shape of the primary rainbow, and noted that they sufficed for 'exploring things beyond our comprehension and delivering ourselves to the derision of the more exacting critics of a physics of causes'. In the mathematical lectures he delivered several years earlier, he considered the study of colors as an example for those subjects of natural philosophy that were particularly remote from certain knowledge, and with respect to which 'there are almost as many different Conceptions and Explications…as there are Authors an Interpreters. Consequently it proves to be difficult to determine or demonstrate anything true about ambiguous and undefined matters of this sort'.[48] Barrow's reflections on the methodological foundations of the natural philosophy of light and colors affirmed those of Boyle. Barrow was content with confining his work primarily to geometrical aspects of this subject matter, unlike Boyle, and later Newton, who chose to investigate it experimentally.[49]

When Newton first published the results of his investigations on the refraction of light and 'the celebrated *Phaenomena* of *Colours*' in 1672, he reiterated the distinction Boyle and Barrow had earlier drawn, and applied it to distinguish his own achievement. 'For what I shall tell you', he proclaimed, 'is not an Hypothesis but most rigid consequence, not conjectured by barely inferring 'tis thus because not otherwise or because it satisfies all phaenomena (the Philosophers universall Topick,) but evince by y[e] mediation of experiments concluding directly & wthout any suspicion of doubt'.[50] 'Hypotheses' as Newton would profess several decades later in his *Opticks*, 'are not to be regarded in experimental Philosophy'.[51]

---

[47] For a detailed study of Barrow's philosophical account of light, see A. Shapiro, 1990.

[48] Quoted from A. Shapiro 1990, p. 168-9.

[49] For Newton's debt to Boyle, see also Mamiani 1986, pp. 17-25.

[50] Newton 1959, i, pp. 96-97. For the published version of Newton's article, see Newton 1672. For the differences between the two versions, see Cohen 1958.

[51] Newton 1952, pp. 404-405.

Boyle's *Experiments and Considerations* recommended an experimental inquiry into the origin of colors. His book, however, was programmatic and did not offer clear solutions to philosophical problems on which students could model their own inquiries. It did not provide Newton with definite instructions or guidelines for advancing his research beyond the initial steps registered in his early notebook. A comparison between Boyle's approach to the experimental study on colors and Descartes experimental reports in his *Dioptrics* and *Meteorology* provided some direction. In Descartes' writings, experimental reports were intended to represent empirical regularities. For example, the order of spectral colors, unequal refraction, or the sufficiency of a single refraction at minimum deviation represented aspects of law-like occurrences. Once the natural regularity was revealed with sufficient clarity, philosophical discourse could then offer a definitive explanation in terms of its underlying nature and the processes that brought a particular nature to generate observable regularity.

Boyle, of course, was also interested in the law-like behavior of the natural world. His experimental writings nevertheless highlighted the work-like characteristics of experiments. Every experiment involved a technique. By means of experiment, specific effects could be generated. Experiments extended and embodied the instrumental actions of the agent that designed and performed them. For example, the exertion of pressure on the eyeball – the simple trial Locke and Newton performed after Boyle – did not perhaps reveal law-like phenomena, though it presented a useful technique for demonstrating that exposure to light was not a necessary condition for generating a sensation of color. Experiments worked. The know-how they required conveyed the experimenter's understanding of those aspects of the natural world that enabled experiments to work in one way rather than another. As I have shown in the previous chapter, this aspect of experimental research was prominent throughout Gray's career. As will be shown in what follows, it dominated Newton's optical studies since 1666.

Newton's report of unequal refraction in his *Quaestiones* echoed Descartes' law-like representation of the phenomenon of spectral colors. The geometrical analysis highlighted the notion that evidence conveyed a natural regularity. Newton noted in addition that 'ye slowly moved ray *being separated* from ye swift ones by refraction, there ariset 2 kinds of colours viz: from ye slow ones blew... From ye swift ones red'.[52] Experimentation

[52] McGuirew and Tamny 1983, p. 432-4; italics added.

with prisms conveyed a technology with a twofold effect, namely the refraction of light and separation between rays. The empirical regularity of refraction acquired its precise meaning in the context of the geometrical study of the interception of light with a denser medium of glass. The effect of separation evoked a more practical image of the use of a prism as a means to manipulate light rays. His early notebook did not distinguish between the two meanings, which, in this context, were both intimately related and highly complementary to one another. The geometrical idiom reigned over the account, and Newton did not ponder the effects of separation. He was not particularly concerned with these effect, because he followed in the footsteps of Descartes and sought to relate the geometrical account of the thread experiment to its philosophical interpretation. Thus, as long as his philosophical studies followed Descartes' example, the manipulative character of the prism was not particularly relevant.

By contrast, in his *Of colours*, Newton renounced the philosophical interpretation and concomitantly seized upon the efficacy of the prism to manipulate chromatic rays of light. His entire manuscript reflected strategic decisions to examine how manipulative the prism was with respect to these rays. It included reports of several dozens of experiments, each of which displayed the effects of refraction by separation under specific conditions that Newton systematically modified from one experiment to another.

The first series of experiments expanded upon his earlier study involving colored surfaces. This was followed by experiments on the refraction of sunlight in a darkened room, as Boyle had suggested. This setting enabled Newton to carefully define and monitor his work space. Beams of light penetrated the room through a small aperture, and the colored images following refraction were projected on an opposite wall. Newton carefully noted the position of the prism, measured the distance between the aperture and the prism, as well as between the prism and the projected images, noting the exact location of the projected images. He then employed a rectangular glass vessel filled with water as a medium of refraction, and later combined it with a prism.

In another series of experiments he employed two prisms, and later added a third. He used these prisms to obtain simultaneous refractions, and then sequenced ones, where one prism refracted the light that was already refracted by the other. In some cases the prisms were attached to one another, and in other cases they were positioned at a distance from one another. Further experiments were performed by rotating one of the prisms around its axis. Newton in addition changed his own position: in some cases he observed the color images from a side, and in others he observed them through the prism.

The wealth of empirical evidence suggested not only a correlation between colors and indexes of refraction, but also a correlation between these variables and the existence of separable rays. Unequal refraction implied separation. Geometry dictated the regularity; yet the prism was the instrument that did the work. It diverted rays from their normal trajectory and separated between them. Moreover, since the prism presumably could effect a separation only because the rays were by nature separable, Newton concluded that the beam of light was composed of rays of specific properties. He began to acquire techniques and practical experience necessary for demonstrating how discrete and distinguishable these rays were. Newton thus modified the meaning of light rays once he began to systematically use the prism as a technique for their manipulation. As Hall pointed out, 'a ray of light, previously a convenient fiction of geometrical optics, has become for Newton a physical entity, though one whose exact definition he was always reluctant to refine'.[53]

The conventionally fictitious notion of ray originated in the traditional division of labor between mixed-mathematics and natural philosophy. The former discipline was devoted to the geometrical analysis of illumination and visualization. The latter was concerned with illumination or visualization as physical processes that characterized the nature of light.[54] Descartes accordingly explained in his *Dioptrics* that since 'the light of a luminous body must be regarded as...its action, you must think of the rays of light as nothing other than the lines along which this action tends'. Though the 'action' of light was of a definite magnitude, it followed from basic principles of geometry that 'there is an infinity of such rays which come from all the points of a luminous body towards all the points of the bodies it illuminates'.[55] Hooke similarly noted that rays were mathematical lines that marked out arbitrary sections within the radius of the propagation of light from its source. Adapting this convention to his

---

53 Hall 1996.

54 For the division between the theory of ray and the philosophical explanations of colors in antiquity, see for example, Smith 1999; and in the Middle Ages, see Wallace 1959; Crombie 1967; Lindberg 1976.

55 Descartes 1985, i, p. 155.

wave mechanics, he represented these geometrical sections as forming the front-line area of a diffusing pulse.[56]

In the late 1660s, Newton's rays were by no means a convention, though they were no less imaginative. In virtue of the function of the prism as a mixed-mathematical instrument, Newton could claim that the chromatic rays of the refracted light beam were an extension of the rays from the incident beam. Since the prism offered, in addition, a technique for separating chromatic rays, he concluded that the separated rays already composed the incident beam. This explained the formation of the spectrum of colors. The incident beam that propagated from the source of light was heterogeneous, and for this reason, none of its separable components could generate a specific color-image. Once the prism separated between rays, their capacity to generate specific color-images that together formed the phenomenon of the spectrum could be revealed. Thus, the new concept of light rays related incident beams to refracted beams in a manner that enabled Newton to explain how spectral colors were formed. The concept was an explanatory construct and as such provided Newton with an alternative to the Cartesian hypothesis that he had adopted in his early notebook.

Newton could still interpret the process of illumination by that hypothesis or one similar to it. In a letter he sent to Boyle six years after he published his new theory of light and colors, for example, he suggested that the universe was permeated by a substance similar to the air but considerably more elastic and 'subtile'. He believed that this substance was responsible for the phenomena of gravity, electricity, fermentation and light. He noted, more particularly, that liquids permeated this efficacious matter and functioned as buffers of light; they differed from one another with respect to their 'density'.[57] During the late 1660s, however, he was determined to set such philosophical interpretations aside and demarcate them from the experimental explanation of colors. His approach enabled him to assume an agnostic position with respect to the philosophical definition of the nature of his newly invented rays, and at the same time to refine their explanatory value and their experimental manipulation with greater precision and scope.

The experimental reports in his *Of colours* explained the various meanings Newton assigned to unequal refractions of light as his research

---

56 Hooke 1961, p. 57.

57 Newton's letter to Boyle, February 28, 1678/9, in Boyle 1672, i, pp. cxii-cxvii; Newton 1959, ii, pp. 288-296.

progressed. In hindsight, the analogy between these meanings and the meanings of electrical conductivity in Gray's writings may be noted. Unequal refraction denoted a fact, which he often represented by means of available geometrical tools; it denoted an experimental technique, which he learned to apply to a wide variety of research projects; and it became the principal tenet of theoretical expositions in his publications.

During the late 1660s, Newton learned to bring the various meanings of unequal refraction to bear upon one another in ways that reinforced the credibility of his studies. In one of his first reports on the simulation of the rainbow, he situated the prism sufficiently far from the wall on which the color-images were projected to amplify the oblong shape of the spectrum. The result was a highly apparent discrepancy between that shape and the shape of a non-refracted beam. The former, he noted, 'should have beene in a round circle' proportional to the hole through which the incident beam of sunlight traversed' (Figure 3.5 on next page).[58] He then applied the received law of refraction to state the fact of unequal refraction, which in turn explained how the discrepancy happened.

In another series of experiments he reported to have held a second prism in the projected spectrum five or six yards from the first prism. The first projection produced the expected oblong shape. The second prism was then employed to demonstrate that color-images remained the same but were further separated on the wall from one another in accordance with the specific refrangibility of rays.[59] The results were predictable, but this technique of double refraction rendered his basic explanatory construct more robust.

Newton chose both of these experiments to demonstrate his new theory in his first publication. He apparently believed that they were among the finest examples of unequal refraction of chromatic rays. The technique which he gradually learned to master could not, of course, demonstrate that his theory was more certain than the hypotheses proposed by other philosophers. It nevertheless demonstrated a new mode of philosophical explanation. Techniques are more or less useful, and their value as such cannot be determined by a theory of knowledge. Ironically, these two experiments later provided the basis for Mach's aforementioned reconstruction of Newton's contribution to science as knowledge.

---

58 McGuire and Tamny 1983, pp. 467-8.

59 Ibid., p. 478-9.

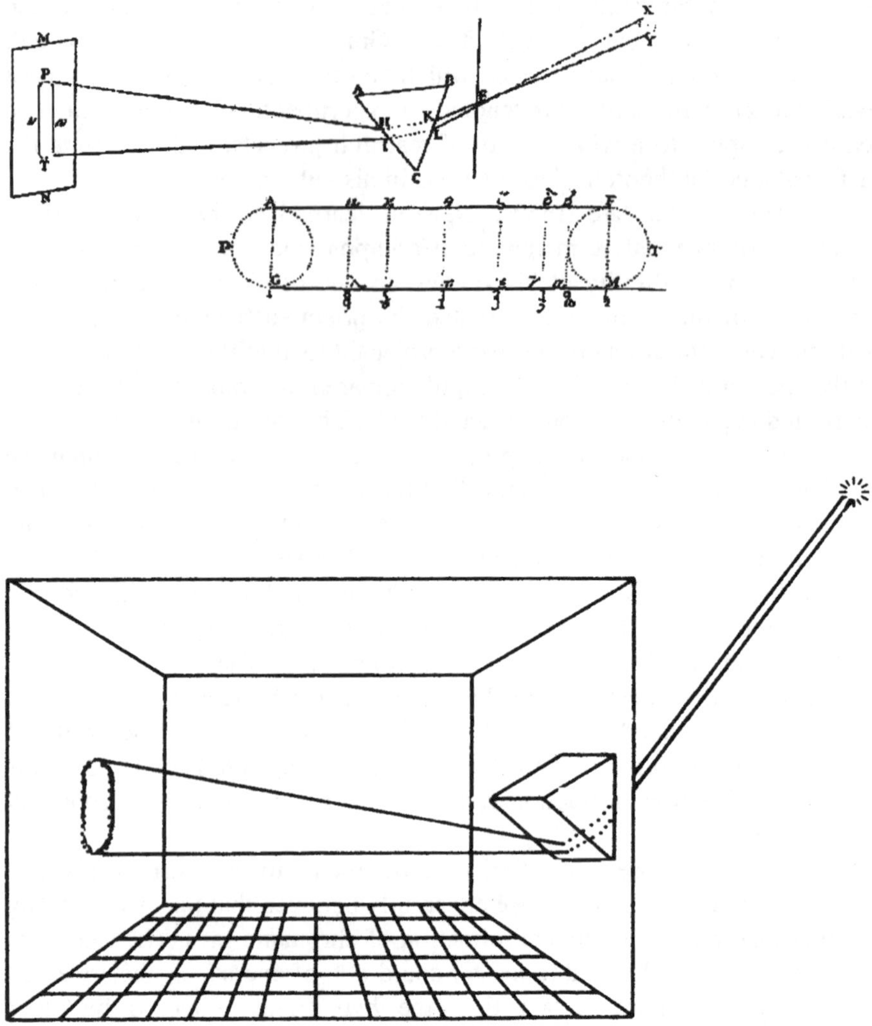

3.5 A reconstruction of Newton's illustrations of the oblong shape of the spectrum, following the refraction of a beam of sunlight with a circular cross-section (based on Newton 1952, 1984)

The burgeoning industry of optical instruments provided a particularly suitable setting for demonstrating the threefold meaning of unequal refraction and its value. The industry was traditionally dominated by the mixed-mathematical study of light, rather than the philosophical discourse on its nature. As noted above, Descartes decided to compromise the quest for certainty to render his philosophy more relevant to this practical subject. Newton's new mode of explanation was more compatible with the practical study of illumination. The unequal refraction of chromatic rays was intended to blur the distinction between theory and technique, or law-like regularities and the effects of work. By the late 1660s he became acquainted with the *Philosophical Transactions* of the Royal Society, which frequently included discussions on the design of optical instruments. At this time, Newton studied James Gregory's *Optica promota*, which included a detailed presentation of the author's invention of a reflecting telescope. Reflecting devices could circumvent the problem of chromatic aberration in refractive lenses. Newton began to realize the relevance of his new theory to the comparative assessment of different designs.[60]

Newton sought to demonstrate how his new explanation could solve both philosophical and practical problems. Solutions to the former could lend credibility to solutions to the latter, and vice versa. In 1669, he commenced his optical lectures at Cambridge with a brief review of the implications of geometrical optics to the improvement of telescopes. He then immediately turned to the systematic presentation of his new theory: Rays of light 'differ from one another with respect to the quantity of refraction: Of those rays that all have the same angle of incidence, some will have an angle of refraction somewhat larger than others'.[61] As will be shown in the next chapter, Newton emphasized the practical value of his novel explanation of colors in his correspondence with the Royal Society. Contemporary philosophers, however, did not fail to recognize that his approach to the study of colors was incongruent with the customs of discourse on the nature of light, despite how practically useful it seemed to be.

---

60 Hiscock 1937; Kuhn 1958; Westfall 1980, pp. 161-3; Newton 1984, pp. 13-4; Schaffer 1989; Simon and Hinkley 2001.

61 Newton 1984, pp. 47-9.

Chapter Four

The Public Transaction of Newton's Optical Research

Introduction

The previous chapter has shown that during the second half of the 1660s, Newton developed a research project on light and colors that transformed the traditional mode of explanation in natural philosophy. Peripatetic and mechanical philosophers commonly searched for a definition of the nature of light, albeit in radically different ways. They commonly aimed to demonstrate that their respective definitions provided the adequate basis for the causal explanation of the modification of light and the phenomena of colors. Newton discarded the problem of nature, and sought to explain this phenomena by correlating its properties with the empirical properties of a beam of sunlight.

He concomitantly transformed the function of experiment in natural philosophical research. Philosophers traditionally conducted experiments that revealed empirical regularity, which served as the explicandum for inquiry into particular natures. They commonly acknowledged, accordingly, that different and even contradictory definitions of a particular nature could provide satisfactory explanations for a particular empirical regularity. In Newton's studies, by contrast, experiments were the means to develop a technique that could show how empirical properties were correlated. The experimental reproduction of a particular correlation demonstrated its regularity. Since this was precisely the aim of his theory, his experimental work and his theoretical studies complemented and reinforced one another.

The subject matter of Newton's studies – light and spectral colors – endowed his innovative approach with potentially enormous significance. The study of light was traditionally at the crossroads of mathematics and natural philosophy. During the 17th century, it supported the new industry of visual and imaging instruments. Moreover, it became one of the main frames of reference for debates about fundamental issues in natural philosophy that accompanied the rise of mechanical philosophy. His short treatise *Of Colours* and the series of professorial lectures he prepared in the late 1660s could thus potentially restructure and redirect one of the leading sciences of the period. Both of these documents, however, were largely in his private possession. Before 1672, Newton was barely known to scholars outside of Cambridge.

It was therefore his first publication, which he originally communicated in a letter to the secretary of the Royal Society, Henry Oldenburg, that initially demonstrated how significant his research actually was. The public transaction of his novel work immediately gave rise to one of the stormiest controversies in 17th century natural philosophy. It lasted several decades, and initially engulfed a group of highly reputed scholars in England and on the continent who by and large rejected Newton's contributions. The controversy – especially in its initial phase – apparently became the most bitter episode of Newton's career.

From the perspective on science as knowledge, the controversy should not have occurred. If Newton's contributions, as Westfall noted, were highly exceptional manifestations of 'the power of experimental enquiry animated by a powerful imagination and controlled by rigorous logic', then the finest scholars of Europe at the time should have embraced them. The scholars, however, consistently disputed their meaning and value. Westfall's appraisal implies a general standard for relating theory and experiment. The standard apparently was not common in the 17th century.

Ironically, the historical study of the reception of Newton's theory became a battlefield of competing philosophical definitions of knowledge, each of which suggested a solution to the problem of theory-choice. Philosophical disagreements notwithstanding, commentators have commonly assumed that a proper definition conveyed universal methodological criteria for assessing Newton's contribution to scientific knowledge. Their assumption implied that Newton's critics, or alternatively, Newton himself, failed to recognize and assimilate the appropriate criteria and thereby made irrational decisions.

Historians who believed that Newton's theory met the criteria for scientific knowledge suggested, accordingly, that his critics failed to understand it, or were unwilling to admit that it genuinely advanced scientific knowledge. Westfall, for example, explained that 'the mechanical philosophy with its demand for picturable explanatory images obstructed the understanding of Newton's discovery that light is heterogeneous'. The critics' 'inability to recognize the force of his demonstrations quickly drove Newton to distraction'.[1]

Other scholars claimed, in contrast, that Newton failed to recognize that he did not adequately prove his theory. Newton was dogmatically loyal to his own ideas, and dogmatically refused to acknowledge that other theories were as valid as his. Bechler thus suggested that Newton's position throughout the controversy expressed 'his utter incapacity to grasp the 'fine structure' of the difference between fact, or phenomena, and theory in those delicate situations where they seemed to fade and blend into each other. These categories were hopelessly mingled in his mind, which fusion, though a handicap in his disputes, was an integral part of his genius'.[2]

From the perspective on science as causal explanation, the controversy appears to have involved a collision between the two modes of explanation. Explanations are more or less useful to different individuals for different reasons. In virtue of the partial subjectivity of explanations, neither Newton nor his critics could be reasonably expected to take an impartial or impersonal stand with respect to the meaning or value of the theories they discussed.

Newton regarded his first publication as his first opportunity to demonstrate to other scholars his new approach to the philosophical study of light and colors. Some of these scholars earned their reputation by demonstrating their own contributions to the philosophical study of the nature of light. Given their vested interest in this tradition, they were expectedly not inclined to endorse a work that undermined its value. This applied in particular to mechanical philosophers, whose contributions were widely recognized at that time to have successfully pioneered a new perspective on the nature of things.

---

1 Westfall 1980, pp. 249, 239. See also Westfall 1962a, 1966; Gillispie 1960, p. 125; Hall and Hall 1962b; Lohne 1968; Laymon 1978; A. Shapiro 1996. For a critique of philosophical reconstructions of Newton's demonstration from the perspective of social studies of science, see Schaffer 1989. A framework for the analysis of the dynamics of the controversy which combines the theory of knowledge with social and psychological dimensions is proposed in Mamiani 2000.

2 Bechler 1974, p. 120. See also Sabra 1967, pp. 231-50, 273-97; Feyerabend 1970.

Newton's first publication thus initiated a process of institutional change in the public transaction of philosophical explanations. This chapter examines how Newton initially sought to mobilize other scholars to support his initiative, and how his early critics undermined his endeavors. The first two sections focus on the rhetorical resources and strategies Newton deployed in his communication with the Royal Society.[3] The first examines the historical and cultural context in which Newton found the resources he needed. The second focuses on the rhetorical structure of his first publication. The final section examines how the collision between the two modes of philosophical explanation structured the controversy between Newton and his early critics.

The Origins of the Rhetorical Structure of Newton's Article

Founded in 1665, the Royal Society's *Philosophical Transactions* joined initiatives in the 'republic of letters' to expedite communication among scholars by the publication of relatively short research works. The Academia del Cimento of Florence began to print its proceedings in 1657, and in January 1665 the *Journal des Sçavans* first appeared in Paris. Three months later, *Philosophical Transactions* was issued in London. Its first editor was Henry Oldenburg, secretary of the Royal Society and one of its founding fellows. As Kuhn noted, Newton 'was the first to advance through this new medium an experimentally based proposal for the radical reform of a scientific theory, and his proposal was the first to arouse international discussion and debate within the columns of a scientific journal'.[4]

The disputes between Newton and his early critics contrasted sharply with the public image of the Royal Society and its journal. Founded in 1660 by the charter of the newly restored monarchy of Charles II, the Society was unanimously portrayed by its founders and fellows as a close ally of the new regime in its firm commitment to restore public peace. Experimental knowledge in particular was advocated as a highly effective remedy for ideological divisions that were commonly attributed to sectarian

---

3 For recent studies in history of scientific writing and the rhetorical styles of scientific argumentation, see Benjamin et al 1987; Nelson et al, 1987; Peterfreund 1990; Dear 1991.

4 Kuhn 1958, p. 28.

doctrines. In the Society's quasi-official manifesto, *The History of the Royal Society of London* (1667), author and fellow Thomas Sprat declared that 'the reason of mens contemning all Jurisdiction and Power, proceeds from their Idolizing their own Wit:... they set up their own Opinions, and worship them. But this vain Idolatry would inevitably fall before Experimental Knowledge'.[5]

Historians have demonstrated that the writings of leading fellows of the Society and their public meetings were intended to display a commitment to codes of public discourse that were closely associated with experimental research. These codes were predominantly designed to secure consensus on matters of fact. They encouraged attention to detailed circumstantial evidence concerning experimental work, and certified degrees and measures of credible witnessing. They carefully distinguished between empirical evidence that was accessible to any reliable observer on the one hand, and the expressions of learned authority individual philosophers chose to display on the other.[6]

Newton's strategic decision to demarcate his work from hypotheses about the nature of things fitted in with the Society's commitment to a separation between the public transaction of factual information on the one hand, and personal opinions, philosophical theories, or sectarian doctrines on the other. His work as a whole, however, deliberately abrogated the distinction between impersonal evidence and authoritative explanation. Newton neither sought to take part in the gentlemanly exchange of impartial empirical information, nor requested the fellows' philosophical interpretation of his experimental findings. He sought to establish a consensus on the authority of his research as a whole. He was determined to communicate to the Society an experimental report that was supposed to lead witnesses to affirm his philosophical conclusions. He therefore needed rhetorical resources to establish a relation between author and disciples which would evoke trust in his personal ability to lead witnesses to a truth.[7]

In 17th century England, the most common source of intellectual authority was the 'word of truth' as it was delivered by a great number of

---

5 Sprat 1667, p. 430.

6 B. Shapiro 1983, 1986; Shapin 1984, 1994; Shapin and Schaffer 1985; Dear 1985; Daston 1988, 1991.

7 Shapin 1994 highlights the importance of gentlemen's codes of trustworthy conduct in natural philosophical discourse in 17th century England. Newton's article appears to be an exception to this custom.

preachers and lecturers throughout the country to a multitude of believers of different social classes and positions.[8] In contrast to the disinterested report on empirical knowledge, the sermon was regarded as a communicative framework in which it was not only expected of the preacher to instill belief, it was his duty to do so. The historical evidence presented in what follows does not directly show that Newton followed the rhetorical style of sermons. It nevertheless sheds light on the historical and cultural context in which the young scholar sought the means to persuade his readers that his quest for the truth was righteous.

During the 17th century, the Protestant clergy of England gradually developed a sermonical style that aimed to rebut the allegedly corrupt authority of scholastic theology and promote literate piety that could be applied to the everyday life of believers. By the 1660s, 'Doctrine' and 'Use' were commonly known as the two pillars that rendered the sermon an effective tool for the discipline of faithful conduct. Preaching to the clergy on his diocese, Thomas Sprat attributed the new code of preaching to the origins of Christianity: Christ, he stressed, 'commanded his *Disciples* not to believe him, but the *Works* that he did'.[9] The ideal sermon was 'plaine and practical', John Evelyn noted in his diary, 'of which sort', he added 'this *Nation* nor any other ever had greater plenty...since the *Apostles* time'.[10]

The Restoration style of preaching was based on earlier reforms which formed an integral part of the evolution of the Protestant movement. The late medieval church regarded preaching as a major clerical duty, and manuals taught the clergy the art of preaching. It was nevertheless commonly assumed that ordinary persons did not need to be able to articulate their faith in an educated manner. They could be implicitly faithful. As Cameron has pointed out, the late medieval notion of 'implicit faith' marked out the division of labor in Christian society between the educated priests and monks who were responsible for articulating the body of Christian knowledge on the one hand, and the lay public who was expected to rely on the educated elite and conduct its secular and religious affairs in pious ignorance of intricate theological matters on the other.[11]

---

8 Fraser Mitchell 1962.

9 Sprat 1696, p. 352; see Patrick 1669, pp. 7-10. For studies relating the new sermonical code to contemporary innovations in prose style, see Jones 1930, 1965.

10 Evelyn 1955, iv, p. 330.

11 Cameron 1992.

Protestantism sought to eradicate this division by infusing religious literacy into the life-cycle of all believers. Godly learning constituted the discipline of religious regeneration. The Protestant sermon thus became the institutional core of the reformed community. The scholastically trained clergy were accused of impressing their lay audience with the frequent use of literary affectations, fanciful conceits, metaphors, and displays of Greek and Latin eloquence. The reformed preachers, in contrast, were called to combine piety with learning to regenerate common notions of faithfulness.

In his *Institutes of the Christian Religion*, Calvin taught that faith was often abused by two different modes of intellectual misconduct. The first consisted of the failure to moderate the 'desire of knowing more than is given to us to know'. Scholars sought to highlight their intellectual talents, and thereby 'involve themselves in hurtful speculations'. Arrogant scholars were 'puffed up with vain science', and could not leave 'a corner of heaven untouched by their speculations'. The results were 'errors, slips, and falls'.[12] The second type of intellectual misconduct was manifested by those who failed to seek divine wisdom because they confined themselves to common knowledge. Calvin accordingly insisted that, 'by knowledge we do not mean comprehension, such as that which we have of things falling under human sense'. Faith involved intellectual effort because 'the human mind must far surpass and go beyond itself in order to reach it'.[13] Faithful learning was predicated on the believer's confidence in God's benevolent intention to reveal His rule in the scriptures and in everyday life.

By the turn of the 17th century, reformed clergymen in England considered scriptural interpretation or 'exegesis' as the core of their special vocation. Puritan reformers in particular realized that the diffusion of ideas could not be left to established incumbents. It was necessary to train new cadres of preachers. Preachers often met in conferences of 'prophesying' that were designed to train preachers, and expedite the 'spiritual' conversion of their communities. Many of them only found a living in the Church by becoming 'Lecturers' in parish churches, paid by the congregation or by some patron with Puritan sentiments. The sermonical literature often established a close connection between the Puritan preacher and the local grammar school.[14] During his visit to England in the early

12 Calvin 1962, i, iii.xxv.10-11, pp. 274-275.

13 Ibid., i, ii.ii.14, p. 482.

14 Simon, J. 1966, p. 381.

1640s, John Amos Comenius, Czech reformer of education, noted that 'a large number of the men and youth copy out sermons with their pens'.[15]

The 'spiritual', 'prophesying', or 'experimental' preacher became the spearhead of the reform movement. Leading preachers commonly recognized, however, that their vocation inherently involved a major predicament. On the one hand, they knew that their ability to appear persuasive hinged on their professional training. On the other hand, the elaboration of preaching techniques rendered the sermon a work of art rather than an inspired act of speaking forth the word of God. If 'prophesy' was a human art of discourse, it obviously stood squarely against the divine message.

The hermeneutic dilemma posed a professional challenge before Puritan divines. They stressed that learning was an indispensable requirement of reformed souls. By propagating ignorance, illiterate preachers advanced idolatry. As Puritan leader William Perkins professed, the Puritan attack on scholastic theology did not by any means undermine the importance of education. 'If any man thinke that by this meanes barbarism should be brought into pulpit; he must understand that the Minister may, yea & must privately use at his libertie the arts, philosophy, and varietie of reading, whilest he is in framing his sermon'. In accordance with Calvin's teachings, Perkins nevertheless emphasized that scholarship was not an end in itself. Its value hinged on the ability of the preacher to turn it into a means of spiritual regeneration. The professional preacher used his learning to demonstrate how the scriptures speak for themselves. 'He ought in publike to conceale all these [sources of knowledge] from the people, and not to make the least ostentation. *Artis etiam est celare artem: it is also a point of Art to conceale Art'.*[16]

The authority of preachers was supposed to be embedded in the text. The 'Word' could in turn become a living guide that made a difference in people's life. Following Perkins, John Geree, a renowned preacher from Saint Albans, explained that 'preaching best wherein was most of God, least of man'. The preacher was devoted to 'the demonstration of Gods power and spirit'. The proper sermon was concomitantly predicated on making and showing the distinction 'between studied plainesse and negligent rudenesse'.[17] According to Richard Baxter, it was important to

15 Quoted from Young 1971, p. 65.

16 Perkins 1609, vol. ii, p. 759.

17 Geree 1646, p. 2.

confine the sermon to an 'evident declaration of the meaning of the text'. If learned authority was merely glossed on the scriptures, preaching became like 'Painted Glass in the Windows that keep out the Light'.[18]

The sermonical style of plain rhetoric was highly compatible with the new codes of philosophical discourse preached at the Royal Society. The former demanded that preachers restrained themselves from speculating about the divine message in public. The latter were designed to demarcate the transaction of experimental knowledge from philosophical opinions.[19] Newton was clearly aware of the affinity between the two cultural codes. In his methodological writings he associated idle speculation with hypothetical explanations, and in his theological writings he associated it with the deliberate distortion of the scriptures. As he pointed out in his *A Treatise on Revelation*, 'they were not the Scribes and Pharisees but the inferiour people who believed in Christ and apprehended the true meaning of his Parables and of the Prophesies in the old Testament concerning him. The wise men of the world are often too much prepossest with their own imaginations and too much intangled in designs for this world'.[20]

The more challenging objective of the plain style of preaching, however, was to embed in the scriptural text doctrinal articles that disclosed a theological truth and offered at the same time a practical guide for faithful conduct. This objective was not commonly associated with the gentlemanly or scholarly exchange of empirical information. In the literature on preaching manuals, the task was often discussed in reference to the distinction between the legibility of the scriptural word on the one hand, and its intelligibility as a sign of divine providence on the other. The art of preaching consisted primarily of the endeavor to identify and demonstrate the sign by analyzing the relations between various scriptural texts.

One of the most popular manuals in Restoration England was John Wilkins' *Ecclesiastes: or, a discourse concerning the gift of preaching*. The manual originally appeared in 1649, and, by the end of the century, was published in eight editions. Wilkins followed the Puritan style, but during the 1660s his *Ecclesiastes* was widely associated with the sermonical style

---

18 Quoted from Simon, I. 1967, p. 15.

19 For detailed studies of the affinity between theological and natural philosophical styles of discourse in 17th century England, see R.S. Jones 1930, 1932, 1965; Fisch 1952; van Leeuwen 1963; B. Shapiro 1974, 1983; Henry 1992.

20 Newton 1974, p. 113.

of Anglican preachers.[21] Wilkins studied at Oxford and was a member of the philosophical circle that included Thomas Willis, Robert Boyle, Christopher Wren, John Wallis, and Seth Ward. These scholars were later among the founding fellows of the Royal Society. Given his wide intellectual interests, Wilkins stressed that the 'gift' of preaching was not only useful to clergymen. Its methodical proposition was 'suitable to the Reason of Mankind' in general, and more specifically was helpful in cultivating the arts and natural philosophy.[22] The next section takes a closer look at Wilkins' manual, and examines how the narrative sequence of Newton's article from 1672 was rhetorically structured topic by topic in accordance with Wilkins' guidelines for the organization of the sermon.

Newton's Gift of Preaching

The gift of preaching, according to Wilkins, was conveyed by 'the right handling and dividing the Word of Truth'. Its presentation was divided into three parts: 'explication, confirmation and application'. The three parts were designed to suit the aim of 'a Divine Orator': 'to teach clearly, convince strongly and to persuade powerfully' (Figure 4.1 on next page).[23]

Explication

The proper sermon began with the exposition of carefully chosen evidence that directly pertained to a particular topic of study and discussion. Its 'Explication', Wilkins suggested, aimed to state 'the true *sense* and meaning of the Subject to be insisted upon' by clear analysis of its 'Terms'.[24] Newton similarly suggested in his *A Short Scheme of True Religion* that the proper 'Rule' for initiating the study of the scriptures was to compare 'the several mystical places of scripture where the same prophetic phrase or type is used, and to fix such a signification to that phrase as agrees best with all the places'. This rule also applied to the study of the physical world. Newton thus noted that he 'received also much light

21 Christensen 1946; B. Shapiro 1969, pp. 75-78.

22 Wilkins 1669, p. 5.

23 Ibid., p. 7.

24 Ibid., pp. 8-9.

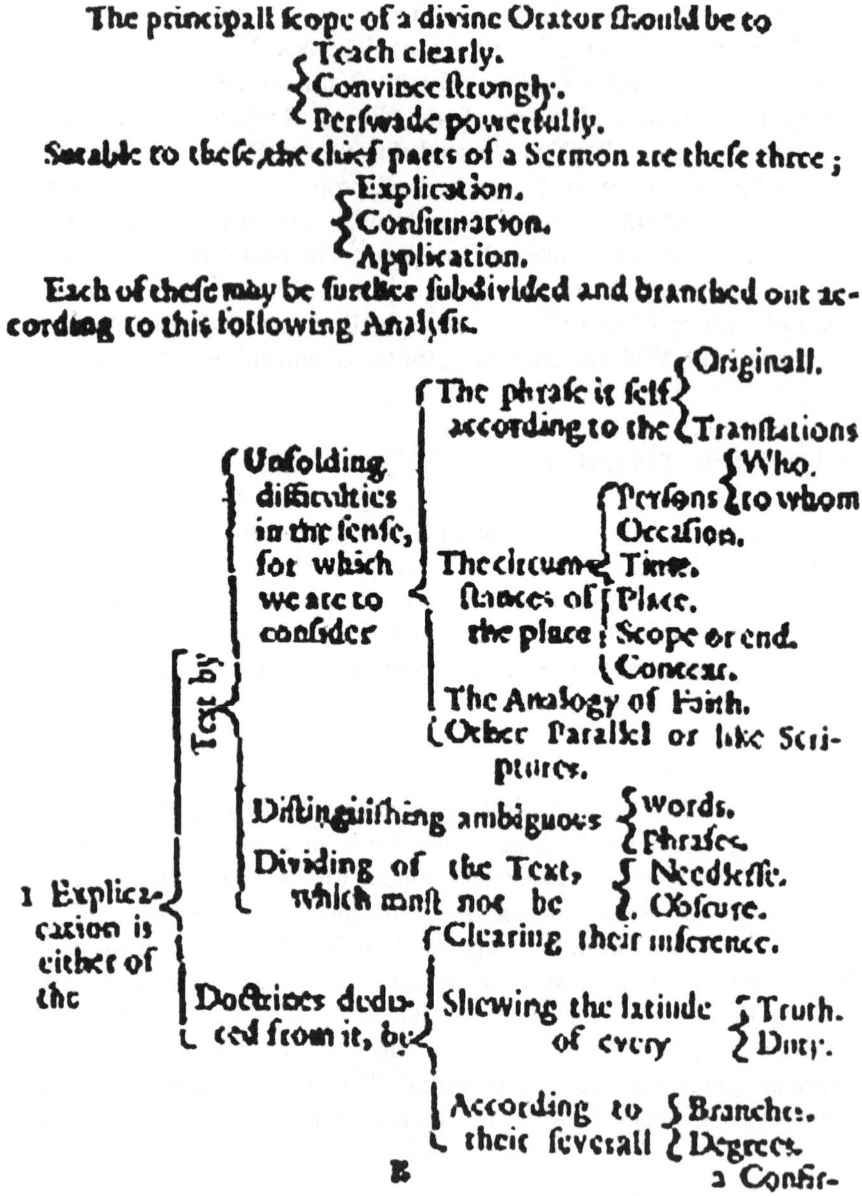

The principall scope of a divine Orator should be to
Teach clearly.
Convince strongly.
Perswade powerfully.
Sutable to these, the chief parts of a Sermon are these three;
Explication.
Confirmation.
Application.
Each of these may be further subdivided and branched out according to this following Analysis.

1 Explication is either of the
Text by
Unfolding difficulties in the sense, for which we are to consider
The phrase it self according to the
Originall.
Translations
The circumstances of the place
Persons
Who.
to whom
Occasion.
Time.
Place.
Scope or end.
Context.
The Analogy of Faith.
Other Parallel or like Scriptures.
Distinguishing ambiguous
words.
Phrases.
Dividing of the Text, which must not be
Needlesse.
Obscure.
Doctrines deduced from it, by
Clearing their inference.
Shewing the latitude of every
Truth.
Duty.
According to their severall
Branches.
Degrees.
B
2 Confir-

4.1 **An extract from Wilkins' diagram of the composition of sermons** (Wilkins 1669, by permission of the Houghton Library, Harvard University)

in this search [for truth] by the analogy between the world natural and the world politic'.

Newton's article from 1672 opened with an autobiographical note that traced his research work to 1666, when he began to study 'the celebrated *Phaenomena* of *Colours*' with a glass prism. The experimental report that followed established the precise 'sense' of the phenomena. He thus at once drew the attention of the reader to the oblong shape of the spectrum and its apparent discrepancy with the received 'laws of refraction'. As noted in the previous chapter, this was one of the first experimental reports in his *Of colours*, which was carefully designed to display geometrical form by distancing the prism from the wall on which the colors were projected.

Newton then proceeded to 'fix' the 'signification' of the oblong shape of the spectrum. He reported to have used various prisms of 'divers thicknesses', changed the size of the hole through which light propagated, and joined two prisms together in order to examine whether the effect produced by the original prism would be neutralized by the second prism. He 'found none of those circumstances material', however, and concluded that ''twas not any contingent irregularity'.[25] Contrasted with the non-refracted light beam shape, the oblong shape was certified as the principal reference to the 'true sense' of spectral colors.

Confirmation

According to Wilkins' instructions, confirmation was preceded by 'the removing' of 'Doubts and Objections'.[26] Once the 'sense' was properly determined by the preacher, it was important to show that other interpretations, which did not accord with the preacher's intention, could not be considered as legitimate options. This rhetorical device was explicitly addressed by Newton in his 'Treatise on Revelation': 'when I had found the necessary significations', he could proceed 'to reject all others as the offspring of luxuriant fancy'.[27]

In his 1672 article, Newton applied the rhetorical device to dissociate his doctrine from mechanical conjectures, particularly those of Descartes. Once the discrepancy between the shape of the spectrum and the

---

25 Newton 1959, i, 92-3.
26 Wilkins 1669, p. 25.
27 Newton 1950, p. 120.

law of refraction was plainly defined and its regularity was confirmed, Newton related that he 'began to suspect, whether the Rays, after their trajection through the Prisme, did not move in curve lines'. His 'suspition' was informed by a recollection of 'a Tennis-ball, struck with an oblique Racket', whose trajectory 'describe such a curve line'. Newton did not mention that Descartes employed the tennis-ball analogy to demonstrate the unequal refraction of color-making rays. Instead, he implicitly suggested that Descartes' legacy amounted to useless speculations. He asserted that he could 'observe no such curvity in them [i.e., the refracted rays]', and confidently rejected the 'fancy'.[28]

According to Wilkins' guide-book, 'confirmation' was established by the analysis of 'the causes or properties, by which a thing is to be defined'. At this stage of his report, Newton introduced his famous '*Experimentum Crucis*', the crux of the demonstration of his new theory. The label of the procedure alluded to Bacon's methodological instructions in his *Novum Organum.*[29] The procedure itself, however, considerably differed from Bacon's proposal with respect to both its methodological and rhetorical functions.

Bacon's methodology of crucial experiments was based on the assumption that the development of philosophical arguments concerning the nature of things was guided by the available empirical evidence as well as by the philosopher's theoretical presuppositions. With the gradual accumulation of empirical evidence, the investigator refined explanations of specific phenomena and separated those which were consistent with the evidence from those which were not. The methodological procedure of 'instantiae crucis' provided the means to render the selection process more systematic. Cross-trials were designed to indicate the 'decisive and judicial' evidence for choosing between two theories, and Bacon accordingly suggested that they ought to disclose new evidence, which would be 'expressly and purposely sought for'.[30]

Hooke followed this procedure in the optical studies he included in his *Micrographia*, and, in a gesture of homage to 'our thrice excellent Verulam', introduced the term '*Experimentum Crucis*', which Newton later reiterated in this exact form.[31] Hooke conducted his crucial experiment in

---

28 Newton 1959, i, p. 94.

29 Ibid., p. 104; Lohne 1968.

30 Bacon 1952, p. 164.

31 Hooke 1961, p. 54; Hesse 1964.

order to discredit Descartes' account of the modification of light in refraction. Hooke claimed more specifically that Descartes did not adequately take into account the material structure of refracting medium. In accordance with his wave analogy, Hooke proposed that the magnitude of the vibration of the refracting medium was proportional to its volume. For any specific medium, colors were most vivid when light was refracted by a medium of a certain volume, and tended to diminish if the volume was either larger or smaller beyond its measurable optimum value. He proposed a crucial experiment with plates of Muscovy glass of different widths, which showed that the brightness of colors depended on the thickness of the refracting glass, and concluded that his own explanation was more consistent with the evidence than Descartes'.[32]

Unlike Hooke and contrary to Bacon's instructions, Newton did not specify the theories that were supposed to be tested by his '*Experimentum Crucis*'. The only theoretical consideration he mentioned up until then pertained to the curved trajectory of a ball; yet its plausibility was already rejected. The crucial experiment was not presented for the purpose of solving Bacon's methodological problem. In accordance with Wilkins' instructions, it was designed to disclose the causal explanation for the oblong shape of the spectrum. Rather then stating his new theory and then examining its empirical adequacy, Newton chose to present it by means of the experimental technique he had earlier perfected.

Newton reported to have employed a second prism, which was held behind a screen on which the oblong shape, produced by the original prism, was disclosed. The chromatic light beam farther propagated through an aperture fixed on the screen. Then, by rotating the first prism around its axis, he separated different parts of the beam. The monochromatic components of the beam reached the screen situated between the two prisms, and propagated through its aperture. The beams were then refracted by the second prism in sequence, in synchrony with the rotation of the first prism. In this order, discrete color-images were projected on the wall (Figure 4.2. on next page).

The report of the crucial experiment disclosed the threefold meaning of unequal refraction, as clarified in the previous chapter. It factually presented the correspondence between colors and specific indexes of refraction. The presentation was, of course, the product of a technique which, in this specific case, enabled him to systematically demonstrate the

32 Hooke 1961, pp. 53-4.

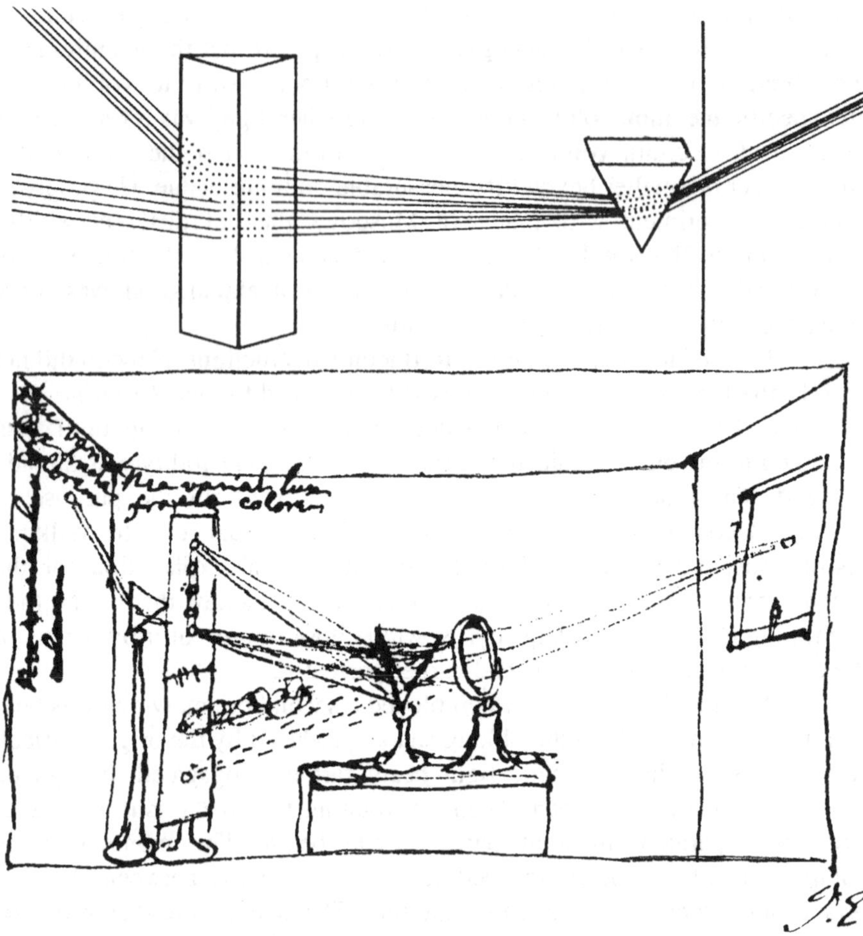

4.2 Newton's illustration of his '*Experimentum Crucis*' with a focusing lens and two prisms (bottom), and a reconstruction of this experiment as presented in his article (top) (Newton 1959, i, p. 108, by permission of the Houghton Library, Harvard University)

correspondence for each color with the gradual rotation of the prism and with the use of a second prism. The report thus presented the technical meaning of unequal refraction. This was one specific application of the technology, which Newton selected from the large repertoire he had been developing since 1666. The theoretical meaning was disclosed in his concluding statement in this presentation: 'The *true cause* of the length of that Image', he explained, 'was detected to be no other, then that *Light* consists of *Rays differently refrangible*, which, without any respect to a difference in their incidence, were according to their degrees of refrangibility, transmitted towards divers parts of the wall'.[33]

The rhetorical structure of Newton's report, then, was precisely the opposite of Bacon's proposal. According to Bacon, the initial stage of the procedure consisted of the presentation of rival theories. Then came the experiment, followed by verdict and choice. Newton began by presenting the fact of the oblong shape of the spectrum. This was followed by an experiment that reproduced this fact and showed, in addition, how it could be technically analyzed. The technical know-how was finally followed by a theory that spelled out its underlying rationale. The technique worked because the 'true cause' of unequal refraction was the heterogeneous composition of light.

The difference in rhetoric reflected, moreover, the difference between two conceptions of the function of experiment. Bacon and Hooke construed of experiment primarily as a means to produce empirical evidence. The procedure was supposed to be neutral with respect to the rival theories. The evidence, rather than the experimental procedure, was the crucial component in the methodology of crucial experiments. In Newton's case, in contrast, both the technical and the theoretical know-how reinforced each other. As shown in the previous chapter, Newton was able to link technology and theory together because his aim was to explain an empirical correlation. Bacon and Hooke, however, were interested in a methodology that could assist in the search for a theory that defined the nature of things. Newton did not address the difference between the two modes of explanation in his first publication. The issue explicitly arose, as will be shown in the next section, only during exchanges with his early critics.

Since Newton's crucial experiment stated in a practical idiom what his causal explanation stated in a theoretical manner, Newton was convinced that the difference between his solution to the philosophical problem of colors differed from, and was better than, solutions that could only offer a hypothesis. The latter, he professed with disapproval, at best

---

33 Newton 1959, i, p. 95.

merely 'satisfies all phaenomena'. This was the 'Philosophers universal Topic'. His explanation, in contrast, was a 'most rigid consequence, not conjectured...but evinced by the mediation of experiments'.[34] Newton apparently regarded the experiment with the two prisms as a crucial trial precisely because its function was not confined to disclosing the evidence his explanation could satisfy. The experiment was crucial because it delineated in a practical manner the setting of a choice. One could either apply the technique for reproducing unequally refracted rays and accept its underlying theoretical rationale, or alternatively, adhere to the received assumption that light was a homogeneous substance.

Application

The 'application' was the main didactic part of the sermon and formed its final section. It systematically summarized the main lessons which could be drawn from the exploratory and analytic parts of the sermon. Wilkins distinguished between two domains of 'application': the 'doctrinal' domain, which included the main articles which the preacher chose to communicate to the public; and the 'practical' domain, in which the preacher explained how the articles of the doctrine were relevant to specific issues of practical concern. With this section of the sermon, Newton concluded his 1672 article.

In the first part of the 'application' Newton addressed some practical problems concerning the improvement of optical instruments. He noted that the problem of crafting focal lenses stemmed from 'the difference of refrangibility' which 'was so great, as I found it'. In designing focal glasses, it became crucial to take into account the correlation between specific rays and specific focal points on the surface of the refracting medium. He then suggested several designs for a reflecting telescope. These designs could 'perfect' optical instruments because they avoided, to a certain extent, the difficulties of aberration which were the inevitable outcome of the fact that light, as 'I told you', 'consists of *Difform* Rays'.[35]

Following the presentation of the practical applications of his experimental research, Newton set out to 'lay down the *Doctrine* first, and then, for its examination, give you an instance or two of the *Experiments*, as a specimen of the rest'. The doctrinal propositions were primarily

---

34 Ibid., p. 97.

35 Ibid., i, p. 96.

designed to expound upon what the '*Experimentum Crucis*' confirmed. There was an 'Analogy 'twixt colours, and refrangibility', and it marked out the '*Original* and *Connate*' property of light-rays.

Newton claimed, in addition, that his theory settled the dispute about the origin of colors. Since colors were innate to the process of illumination, they could not be considered qualities of physical bodies. This was, of course, the main theoretical claim mechanical philosophers made to rebut the scholastic view, as shown in the previous chapter. Newton, however, carefully dissociated his proof from a particular mechanical explication of the nature of light. Alluding to the grammar of peripatetic philosophers, Newton explained that color was the 'quality' of rays, and for this reason one ought 'to call' a ray a 'substance'. The nature of this substance, or what it was made of, were issues for a separate discussion. A report on experimental research was meant to be confined to matters of fact. It was therefore important that it did 'not mingle conjectures with certainties'.[36]

The Dispute with the Critics

Newton's first publication was comprised of a rhetorical edifice of interdependent and complementary facts, experimental techniques, and explanations. The synthesis of these components was the outcome of several years of research work that transformed the relationship between empirical and explanatory practices in the context of the philosophical study of light and colors. On behalf of the Royal Society, Oldenburg brought Newton's article to the attention of Hooke, the Society's curator of experiments, and sent copies to several prominent scholars on the continent for their review.

The scholars read the article in consideration of their well established philosophical approach to the study of light and colors, and the edifice fell apart. They carefully examined the facts, the experiments, and the explanations, but they did not acknowledge their synthesis. Thus began the controversial exchange between Newton and his critics. They commented on the facts, the experiments, or the explanations from their various perspectives, while Newton, time and again, tried in response to

[36] Ibid., p. 100.

reassemble the synthesis and show how its integrated components were mutually supportive.

In Newton's article, statements of fact were integrated into concrete representations of his new theory. However, empirical facts, in contradistinction to the nature of things, were mere accidents. Newton's empirical findings were accordingly construed by the critics as contingent upon the specific experiments he performed. Moreover, the critics did not confine their review to the facts that were presented in his article. They appealed to the evidence they had already obtained before they received Newton's article, or to new empirical studies they conducted on their own in response to Newton's article. The empirical findings and experiments that Newton presented did not have a privileged status. Hooke thus wrote to Oldenburg that the 'many hundreds of tryalls' he had performed in the past were as relevant as Newton's, noting that he did not develop his own explanation 'without first trying some hundreds of expts'.[37] Given his highly proficient experience, he did not intend to examine Newton's findings simply by following the latter's instructions.

Francis Linus, a Jesuit scholar from the English College of Liége, questioned the integrity of Newton's findings. In a letter to Oldenburg from 1674, a year before the aging scholar died, Linus insisted that the oblong shape described in Newton's article could only be observed when sunlight was reflected from clouds and reached the prism with an oblique angle of incidence.[38] Newton briefly replied that the experiments were performed on a clear day, but Linus did not accept Newton's testimony.[39]

A somewhat more collaborative exchange was resumed a couple of years later by Anthony Lucas, Linus' successor at the English College. Lucas initially reported to have carried out a series of experiments to test Newton's theory. The evidence he obtained did not confirm it.[40] Newton, however, refused to discuss Lucas' experiments. He requested that this scholar confined his investigation to the '*Experimentum Crucis*', claiming that 'it is not number of Exp^{ts}, but weight to be regarded; & where one will do, what need of many'?[41] Lucas consented, and later reported that darker colors were indeed more refracted than brighter colors, just as Descartes

---

37 Ibid., i, pp. 110-1.

38 Ibid., i, pp. 317-9; Cohen 1958a, p. 149.

39 Newton 1959, i, pp. 328-9; Cohen 1958a, pp. 150-1.

40 Newton 1959, ii, pp. 8-12.

41 Ibid., p. 79.

and other researchers had observed long before Newton, though he found that the oblong shape was considerably smaller than the one described in Newton's article. He reported, in addition, to have placed the two prisms 'exactly according to Mr. Newtons directions', but nevertheless 'constantly found' that 'violet rayes' were 'accompanied by a considerable quantity of red ones, upon ye paper behind the second prisme'. He pointed out that he had obtained similar evidence by means of other experiments as well.[42]

Lucas' critical response directly addressed one of the most innovative aspects of Newton's contribution – the relation between fact, experiment, and theory. In this context, this relationship appeared to be one of the most vulnerable aspects of Newton's contribution. His new theory implied that it was technically possible to separate light rays in accordance with their 'connate' color and specific refrangibility. This was what the crucial experiment was designed to show. Yet neither Newton nor his critics could differentiate between each and every hue of the spectrum. There was inevitably a discrepancy between the explanatory construct of unequally refrangible rays on the one hand, and the experimental fact of unequally refracted rays on the other. Lucas did not find rays that were unambiguously unequal in refrangibility, and therefore concluded that the explanation was incorrect.

Christiaan Huygens, who was by then a renowned member of the Académie Royale des Sciences, addressed the problem earlier, though from a more philosophical perspective. In his view, the assumption that light was heterogeneously composed of 'so many diversities' turned the explication of the nature of colors extremely cumbersome. Huygens constructively proposed that the indefinite variety of hues could be considered a mixture of two principal colors.[43] His suggestion implied, however, that Newton's theory must be modified.

Newton declined Hyugens' proposal, since the rationale behind his entire experimental project was that all the rays were unequally refrangible. Rather than changing the theory, he suggested to Oldenburg that Huygens would more seriously apply himself to the technical task that he himself had already undertaken. 'Perhaps he [i.e., Huygens] would sooner satisfy himself by resolving light into colours as far as may be done by Art, and then by examining the properties of those colours apart, and afterwards by trying the effects of reconjoyning two or more or all of those...This will

42 Ibid., pp. 104-8.

43 Newton 1959, i, pp. 255-6; Cohen 1958a, p. 136.

prove a tedious & difficult task...but I could not be satisfied till I had gone through it'.[44] Newton thus acknowledged the technical difficulty, but, unlike Huygens, considered it as an integral part of the challenging work that experimental solutions to philosophical problems could raise. The challenging task could be construed, conversely, as a technical complication arising out of Newton's muddled thinking.

When Lucas claimed that the violet light was not separated from the red light, Newton began to understate the factual meaning of unequal refrangibility and highlight its technical meaning. A great deal of technical experience was necessary to disclose the fact in a satisfactory manner. He therefore offered Lucas more detailed experimental instructions. He advised Lucas to divide the project into more manageable tasks. It was important first to examine 'whether some rays be more frangible y^{n} others. What ye colours of ye rays differently refrangible are, or wither they have any appropriate ones belongs to an after enquiry'.[45] Newton concomitantly redefined the role of the '*Experimentum Crucis*', and pointed to Lucas that 'I make no mention of colours while I am prouving different refrangibility by y^{e} *Experimentum Crucis* but after y^{t} begin to tel you y^{t} ye rays wch differ in refrangibility differ also in colour, reserving this to be proved by other Experiments'.[46] He apparently hoped that his critics would learn from his own research experience and follow his lead. Lucas agreed to exchange information with Newton, but similar to Hooke, he refused to consider Newton as an authority on these matters.

Edme Mariotte, a founding member of the Académie Royale des Sciences, also noted that that chromatic rays were not separated as the new theory proposed. Mariotte's earlier investigations on the physiology of plants, human vision, and the pressure of air were widely acclaimed for their empirical precision. His prism experiments resembled Newton's, but he too did not consider Newton's experimental reports as authoritative instructions. Rather than attempting to learn how to reproduce Newton's results from these reports, Mariotte conducted his own experimental investigation.[47]

In 1681, Mariotte published his *De la nature des couleurs*, which included new measurements of primary and secondary rainbows, and

---

44 Ibid., p. 264.
45 Newton 1959, ii, pp. 257-8.
46 Ibid., p. 258
47 Schaffer 1989, pp. 92-3.

systematic experimental simulations of rainbow colors. His evidence, he professed, demonstrated that 'a given portion of light receives different colors as a result of different modifications, and that the ingenious hypothesis of Mr. Newton should not be admitted'.[48] In the context of these exchanges, Newton's strategic decision to dissociate his work from the philosophical discourse on the nature of things seemed to work against him. His new theory was not supported by general philosophical considerations, and its credibility derived exclusively from experiments. Empirical findings that were not consistent with his reported results apparently demonstrated that the theory was simply wrong.

Newton's new mode of explanation raised additional objections. Since philosophers commonly assumed that different theories could provide legitimate explanations for the same phenomena, Newton's critics refused to credit his theory with special authority even if it was consistent with some of the evidence. They construed his article as a contribution to the traditional philosophical transaction of hypothetical explanations, rather than as an attempt to transform it. Thus, in one of the first unsolicited reviews of Newton's article, Ignace Pardies, a professor of mathematics at the Collége de Clermont, acknowledged that the new theory was 'ingenious', but was nevertheless a 'hypothesis'.[49]

Newton politely replied to Pardies that he was 'content that the Reverend Father call my theory an hypothesis if it has not yet been proved to his satisfaction'. His reply, however, appeared to be beside the point, if not arrogant, since his critics assumed that philosophical explanations were hypothetical however empirically adequate they were found to be. Empirical evidence could not on its own prove a claim to the knowledge of causes. In as much as Newton proposed a philosophical explanation, it was and would always remain a hypothesis. This issue was most clearly pressed by Hooke. In his correspondence with Oldenburg, Hooke insisted that Newton's theory must be considered on par with other hypotheses that successfully accounted for the phenomena. 'The same phenomena will be salved by my hypothesis as well as by his without any manner of difficulty or straining'.[50] Newton's experimental report was 'very subtill and

---

[48] Quoted from Guerlac 1981, p. 99.

[49] Pardies to Oldenburg, March 30, 1672, in Newton 1959, i, pp. 130-3; Guerlac 1981, pp. 81-2.

[50] Newton 1959, i, p. 111.

ingenious', Hooke acknowledged, 'but I cannot think it to be the only hypothesis; not soe certain as mathematical demonstrations'.[51]

In 1676, Hooke took part in a committee that was appointed by the Society's fellows to 'try Mr. Newton's experiment controverted by Mr. Linus'.[52] The committee successfully performed the '*Experimentum Crucis*', and several weeks later it was repeated before the fellows 'according to Mr. Newton's directions, and succeeded, as he all along asserted it would do'.[53] Hooke nevertheless continued to insist that the facts neither proved Newton's theory nor refuted the explanation he had proposed in his *Micrographia.* He reaffirmed Huygens' claim that Newton's theory merely complicated the philosophical definition of the nature of light. Newton's theory implied that 'all these motions, or whatever it be that makes colors, should be originally in the simple rays of light'. Hooke could 'conceive the white light or uniform motion of light to be compounded of thousands of compounded motion' within the theoretical framework he espoused, but added, 'I see no necessity of it'.[54]

Hooke further noted that Newton's refusal to admit that his theory was hypothetical violated the ethics of the public transaction of knowledge, which the Royal Society eagerly sought to guard. In a letter which was apparently addressed to Lord Brouncker, the president of the Royal Society, Hooke thus explained that his debate with Newton did not solely pertain to the content of their respective theories. It seemed that Newton transgressed against the 'freedom & liberty of Discoursing and Arguing [which] ought to be Tollerated'. Hooke appealed to Bacon's 'Authority' which 'warranted' that 'the advancement of Philosophy' could only be achieved by 'the examination of hypotheses by experiments & the inquiry into experiments by hypotheses'.[55]

Hooke was correct. Newton did not play by the rules. He was determined to transform them. Newton explicitly stated in his article that he did not approve of the methodological principle of saving the phenomena – 'the Philosophers universall Topick'. His sermonical style implicitly

---

51 Ibid., p. 113.

52 Birch 1757, iii, p. 312.

53 Ibid., p. 313.

54 Newton 1959, i, p. 114.

55 See also the 'memorandum' Hooke composed several months later, in June 1672, in Newton 1959, i, pp. 195-197. For his comments to Brouncker, see Newton 1959, i, pp. 198, 202.

condemned the 'liberty' to propose hypotheses – something Hooke was determined to defend – as corruptive and misleading. Rather than assisting the search for the truth, it led philosophers to impiously mistake their speculative arguments for the order of divine providence. At this stage, however, Newton did not yet spell out in public what the new rules precisely were.

Newton's aversion to hypotheses was neither purely ethical nor methodological, or for that matter, theological. Newton associated hypotheses with the traditional mode of philosophical explanation. The use of hypotheses and their public transaction constituted the old philosophical regime against which his work stood. Yet none of Newton's early critics acknowledged that he presented an alternative. Quite to the contrary, it seemed at least to some critics that Newton simply failed to contribute to the discourse on the nature of light and colors. As Huygens pointed out to Oldenburg shortly after the publication of Newton's article, the latter 'has not taught us, what it is wherein consists the nature and difference of colours, but only this accident (which certainly is very considerable) of their different refrangibility'.[56]

Newton, on his part, openly acknowledged that he was indeed not concerned with this philosophical issue. He wrote to Oldenburg, in reply to Huygens, that 'I never intended to show wherein consists the nature and difference of colors, but only to show that *de facto* they are originall & immutable qualities of the rays wch exhibit them, and to leave to others to explicate by Mechanicall Hypotheses the nature & difference of those qualities'.[57]

This reply to Huygens encapsulated the strategy Newton gradually developed as he tried to steer his work away from the attacks of his critics, and to clarify the rules that demarcated the new regime of the public transaction of philosophical explanations from the old. The philosophical study of the physical world was divided into three levels. The elementary level included phenomena such as the oblong shape of the spectrum. Phenomena could be observed but not necessarily properly understood. In this respect, phenomena were analogous to the legible but not necessarily intelligible words of the scriptures.

---

56 Cohen 1958a, p. 136. For similar assessments of Newton's *Opticks* and his theory of gravitation in the *Principia*, see, respectively, Guerlac 1981, pp. 109-11; Koyré 1965, p. 114-6.

57 Newton 1959, i, p. 264. For his similar reply to Hooke, see p. 174.

The second level included the facts that related the phenomena. Newton's new theory belonged to this level. Facts were representations of phenomena. They were crafted to render phenomena intelligible. A factual representation explained the phenomena, and was made of phenomena. The latter were the resources which the student processed into the former. Newton often used technical terms such as deduction or induction to denote this process. Induction and deduction included, but were not confined to, logical inferences. His new theory, for example, could not be logically inferred from the oblong shape of the spectrum. The transformation of phenomena into factual representations was a creative endeavor, an art that was comparable to the preacher's 'right handling' of the 'Word of Truth'.

The third level included hypotheses which in one way or another related the nature of things. This was the level at which philosophers traditionally plied their trade. Since hypotheses – in contradistinction to factual propositions – were not made of phenomena, they were inherently speculative. Their credibility was accordingly highly questionable. Hypotheses in natural philosophy were analogous, in this respect, to the scholastic teachings of the Catholic Church. They were both like 'Painted Glass in the Windows that keep out the Light'. According to Newton, this level was not a necessary part of the philosophical study of the physical world. Moreover, it could undermine the search for truth if it was not properly demarcated from the second level.

It was within the framework of this scheme that Newton presented experimental philosophy as the new spearhead of the study of the physical world, and distinguished it from traditional natural philosophy. As he succinctly professed in the fourth 'Rule' of 'philosophizing' in the *Principia,* 'in experimental philosophy we are to look upon propositions inferred by general induction from phenomena as accurately or very nearly true, notwithstanding any contrary hypothesis that may be imagined, till such time as other phenomena occur, by which they [i.e., the inferred propositions] be made more accurate, or liable to exceptions. This rule we must follow, that the argument of induction may not be evaded by hypotheses'.[58]

During the exchange with his early critics, Newton began to spell out the various aspects of his conception of the experimental philosophy. In 1672 he wrote to Hooke that he knew that the experimental findings that he had presented in his article could be explicated by different mechanical

58 Quoted from Koyré 1965, pp. 261-72.

hypotheses, including Hooke's. 'And therefore I chose to decline them all, & speake of light in general terms, considering it abstractedly as something or other propagated every way in streight lines from luminous bodies, without determining what that thing is'.[59] 'The best and safest method of philosophizing seems to be, first to enquire diligently into the properties of things, and to establish those properties by experiments and then to proceed more slowly to hypotheses for the explanation of them. For hypotheses should be employed only in explaining the properties of things, but not assumed in determining them'.[60] Thus, the initial failure to persuade his early critics led Newton to pronounce more clearly and consistently how the public transaction of philosophical explanations ought to be regulated.

As the distinction between experimental philosophy and the discourse on nature was clarified, Newton was willing to consider the nature of light in public. In 1675, he decided to present an essay entitled *An Hypothesis explaining the Properties of Light discoursed of in my severall Papers* to the Royal Society.[61] The essay was based on revisions of his previous studies after Descartes which focused on the interaction between light globules and the ether medium. The highly diluted medium was permeated by various 'spirits' that gave rise not only to optical phenomena, but also to electrical and physiological phenomena as well as gravitational motion. As shown in Chapter Two, these ideas later provided the framework for Hauksbee's curatorial investigations.

Newton took this opportunity to explain that he 'had formerly purposed never to write any Hypothesis of light & colours, fearing it might be a means to ingage me in vain disputes'.[62] Newton did not specify precisely when he had made this decision. The analysis in the previous chapter suggests that it was made in 1666, when he embarked upon the experimental investigations that he recorded in *Of colours*. Newton then insisted that his *Hypothesis* should be considered separate from and independent of his experimental investigation: 'This I thought fitt to Expresse, that no man may confound this [i.e., his 'Hypothesis'] with my other discourses, or measure the certainty of one by another, or think me

59 Newton 1959, i, p. 174.

60 Cohen 1958a, p. 106.

61 Newton 1959, i, pp. 362-86.

62 Ibid., p. 361.

oblig'd to answer objections against this script. For I desire to decline being involved in such troublesome & insignificant Disputes'.[63]

Ironically, the dispute between Newton and his critics always centered on his experimental work, rather than on his speculations. From the perspective on the history of science as knowledge, it should have been the other way around. From the perspective on science as explanation, it appears that Newton's experimental research was disputed because it was inseparable from a mode of explanation that breached and undermined the conventions of public transactions of the discourse on nature. Moreover, the specific explanation that Newton proposed contradicted the conventional assumption that the nature of light was uniform.

However, the critics did not dominate institutions of learning. The Royal Society by and large favored the experimental discovery that challenged the supremacy of philosophical speculations. Scholars who were primarily interested in the mathematical study of light and its practical applications were not troubled by the fact that Newton had little to say about the nature of light. Like the former Lucasian professor Barrow, they did not regard the search for a definition of the nature of light and colors as a particularly valuable pursuit. During the early 1670s, John Collins, who was a trader in mathematical works and an amateur mathematician, brought some of Newton's studies to the knowledge of mathematicians in England and Scotland. One of Newton's early disciples was David Gregory, the nephew of James Gregory, who was appointed professor of mathematics at Edinburgh in 1683. During the 1680s, Newton's theory was taught at Scottish universities.[64]

In 1695, Gregory was appointed Savilian professor of astronomy at Oxford, and there he taught that Newton's theory laid the foundation for a unified discipline of optical research that was no longer divided by 'any sect of philosophers'.[65] The message was reiterated in 1701 by John Keill, a student of Gregory's and a lecturer of natural philosophy, who published one of the first Newtonian textbooks.[66] It was officially sanctioned by the Royal Society several years later when Newton was elected its president. When John Desaguliers became the curator of experiments, experimental reproductions of unequally refrangible rays were routinely performed

---

63 Ibid., p. 364.

64 Guerlac 1981, p. 88; A. Shapiro 1996.

65 Quoted from Hiscock 1937, p. 2.

66 Keill 1720, p. 1-3.

before scholars from Holland, France, and Italy who visited the Royal Society. Reflecting on the dispute between Newton and his critics, Desaguliers noted that those who failed the technical task sought 'victory' rather than 'truth'.[67]

[67] Desaguliers 1716, 1722, 1728. On Desaguliers' assistance to Newton, see Heilbron 1983, pp. 90-3; and more broadly on the role of the Royal Society in promoting Newton's optical research, see Schaffer 1989.

Chapter Five

Remodeling Human Understanding: Locke's Laboratory of the Mind

Introduction

The previous chapters have focused on two major discoveries in early modern science – Newton's discovery of the heterogeneous composition of sunlight, and Gray's discovery of electrical conductivity. These discoveries were directly related to one another – both were the result of the initiative to transform the mode of explanation that structured philosophical learning for centuries. These discoveries indicated that the initiative was not confined to specific areas of philosophical research. Newton's conception of experimental philosophy introduced a new regime whose scope was as wide as the collective endeavor to advance human understanding of the physical world.

The initiative engendered a cultural change that centered upon the relationship between explanatory and experimental practice in natural philosophy. Change in these practices tacitly involved cultivating new modes of thought and experience. It could be expected, therefore, that the initiative would be pursued by other scholars and by other means to explicitly address broader issues concerning human learning and its relation to human conduct. The previous chapter has shown that Newton was concerned with some of these issues, especially in reference to the study of the scriptures. He did, however, systematically address them in his published works.

This chapter sets out to precisely assess the scope of experimental philosophy as an agent of cultural change by examining Locke's work, particularly his *An Essay concerning Human Understanding*. Locke's *Essay* was originally published in 1689, and was immediately recognized by a wide audience of supporters and critics in England and on the continent as a highly ambitious endeavor to reform human learning. By the early part of the 18th century, Locke's *Essay* and Newton's *Principia* and

Opticks were together widely regarded as the literary emblem of a new science. It is therefore important to examine how closely Locke's intentions were related to Newton's.

In the late 1680s, Locke prepared a detailed review of the *Principia* for the widely read *Bibliothèque Universelle*, upon the request of its editor Jean LeClerc. Newton and Locke became close friends in the early 1690s.[1] Locke retrospectively acknowledged the relevance of his work to Newton's, and emphasized their complementary endeavors in 'The Epistle to the Reader' that prefaced the first edition of the *Essay*. The 'incomparable Mr. Newton', Locke proclaimed, was among the 'Master-Builders, whose mighty Designs, in advancing the Sciences, will leave lasting Monuments to the Admiration of Posterity'. In the presence of these monuments, Locke could only find, he acknowledged, the 'Ambition enough to be employed as an Under-Labourer in clearing Ground a little, a removing some of the Rubbish, that lies in the way to Knowledge'.[2]

Locke and Newton nevertheless pursued their major studies independently of one another. Locke began to prepare material for his *Essay* in two consecutive manuscripts, now known as 'draft A' and 'draft B', completed in 1671. At that time, Newton's innovative studies were unknown even to Isaac Barrow, who was one of Newton's closest colleagues at Cambridge. Locke composed the final draft of the first edition of his *Essay* during the mid 1680s at his refuge in Holland, but even then he was apparently unfamiliar with Newton's research. If Locke's studies were closely related to Newton's endeavor to transform natural philosophy, then both of them must have been inspired by another scholar whose work had already been familiar to both by the late 1660s. This connection is later explored in Chapter Seven.

In modern narratives of the history of western philosophy, Locke has often been portrayed as one of the founders of modern empiricism. Empiricism has been unanimously considered a philosophical approach to the problem of knowledge, especially scientific knowledge. The central dogma of science studies has thus become associated with Locke's *Essay* – arguably more so than with any other philosophical work of the 17th century. Though Locke's concern over the problem of knowledge was indubitable, even a casual inspection of the *Essay* would indicate that he

---

1 For studies on Locke's acquaintance with Newton and his work, see Axtell 1969; Rogers 1978.

2 Locke 1975, pp. 9-10.

considered the problem as one aspect of a broader and more fundamental subject matter. Its title refers to human understanding rather than knowledge. Of the four books that compose the *Essay*, only portions of the last book systematically examine the problem of knowledge.

Locke's empiricism has been associated with two principal claims, namely that knowledge originates in experience, and that experience is the ultimate authority in the evaluation of claims to knowledge. As historians of science and philosophy have convincingly demonstrated, however, these claims were widely endorsed by scholars long before Locke began to write drafts for his *Essay*.[3] It is highly unlikely that Locke devoted his *Essay* to reiterate a philosophical commonplace. It therefore appears that the central dogma, and in particular the modern notion of empiricism as a theory of knowledge, failed to adequately identify what Locke's *Essay* was primarily about, and what its distinctive contribution intended to achieve.

This chapter begins to examine the *Essay* from the perspective on science as explanation. More specifically, the novelty of Locke's *Essay* is assessed in the context of philosophical discussions of human cognition and intellect as a means to grasp causal relations in the physical world. The first section shows that philosophers commonly sought to relate the study of the human mind to the study of methods of philosophical learning. Explications of the nature of human beings as rational creatures were traditionally presented as partial demonstrations of the capacity of philosophical research to ascertain the nature of other objects. The second section examines how Locke's *Essay* separated the analysis of human understanding from philosophical discourse on the nature of things, and situated it within the framework of the performance of mental tasks within everyday life. Locke concomitantly sought to demonstrate that these tasks were routinely accomplished without the knowledge of particular natures. This chapter then examines how Locke remodeled human understanding within this context, and shows the remarkable affinity between his new model and Newton's experimental philosophy.

---

[3] Both of these claims have been traced to the writings of Aristotle. The authority of experience in the evaluation of theories in the context of late medieval and early modern natural philosophy is discussed, for example, in Crombie 1953; B. Shapiro 1974, 1983, 2000; Hacking 1975; Popkin 1979. Sources on the experiential origins of knowledge are specified in the next section of this chapter.

The Received View on Human Nature and its Capacity to Grasp the Nature of Things

In historical studies on the theory of knowledge, 'empiricism' and 'rationalism' were traditionally portrayed as two major types of theories that were espoused by scholars in the early modern period. Debates between advocates of the rival theories marked out, as Ayers recently suggested, 'one of the great watersheds of early modern epistemology'. Reaffirming the received historical view on Locke as an empiricist, Ayers drew the dividing line in the following manner: 'Gassendi, Hobbes, and Locke all categorically asserted the independent authority of the senses as knowledge-producing faculties...On the other side, Platonistic or Augustinian philosophers such as Mersenne, Descartes, and Arnauld firmly subordinated the senses to the intellect'.[4]

The previous chapters have shown, however, that this dividing line does not adequately apply to 17th century natural philosophy. For example, Descartes studies' of spectral colors highlighted the autonomy of empirical research. Descartes sought to accommodate his general ideas of matter and motion to the results of his experiments and to the geometrical analysis of the phenomena. Conversely, mechanical philosophers commonly acknowledged that their theories were not directly derived from experience. Since they all sought knowledge of the mechanical causes of the phenomena, they could not consistently claim that the senses could be applied to advance this knowledge independent of the intellect.

Theoretical differences notwithstanding, philosophers construed their common search for a definition of the nature of things partly in light of a common view of their capacity as human beings to achieve this goal. The philosophical study of the nature of human cognition and understanding was thus closely related to studies on logic and the methods of learning. These disciplines were widely recognized as mutually interdependent parts of a single enterprise in the early modern period.[5]

The philosophical study of cognition was part of faculty psychology, a science that grew out of scholastic commentaries on ancient texts – most notably Aristotle's *De Anima* – that were concerned with the study of the soul. These studies were not confined to descriptive accounts

---

4 Ayers 1994, p. 49. For studies of Locke's theory of knowledge, see also Woolhouse 1971, 1994; Norton 1980; Kroll 1984; Ayers 1991, i; Tipton 1996.

5 Pansau 1997; Michael 1997; Hatfield 1997.

of the cognitive aspects of human behavior. One of their principal aims was to demonstrate that human beings were naturally endowed with faculties and dispositions necessary for the production of philosophical knowledge.

Faculty psychology was predicated on the assumption that human cognition originated in the animal nature of human agents. Cognition was generally defined as the innate capacity of animals to acquire forms of objects other than themselves. The scholastic approach to the study of human cognition was based on the doctrine that all knowledge originated in the passive reception of forms by sense organs and by the faculties of common sense and imagination. These faculties coordinated the information of the five external senses and conveyed it to intellectual faculties. A reiterated metaphor of the process was the impression of a seal on a piece of wax. The concrete form of the object of cognition was impressed on specific sense organs. The impression formed a unified mental image that resembled the concrete form of the object.[6]

The mechanical philosophers qualified late medieval doctrines by espousing a materialist and non-teleological image of sensory stimuli. The term *idea* became popular among the critics, who associated its meaning with the purely mental contents of thought and perception, in contradistinction to the corporeality of physical bodies.[7] 'Ideas', Descartes explained, 'are formed of the objects which strike the senses'. They should not be mistaken for 'little images flitting through the air, called 'intentional forms', which so exercise the minds of philosophers'. The soul contemplated its own ideas, rather than 'certain images transmitted by objects to the brain'.[8]

Descartes nevertheless adhered to the scholastic doctrine concerning the process of sensory perception. He explained that concrete images of external objects were formed by the sense organs and common sense, located alongside the imagination in the brain gland (designated by H in Figure 5.1 on next page). External senses and common sense operated simultaneously. The former responded to discrete sensory stimuli, and the latter 'functions like a seal, fashioning in the phantasy or imagination, as if in wax, the same figures or ideas which come, pure and without body, from the external senses'.[9] The idea of an arrow, for example, consisted of a

6 Brennan 1941, pp. 111-46; Pasnau 1997.

7 Ayers 1991, i, pp. 20-1.

8 Descartes 1985, i, pp. 153-4, 165.

9 Descartes 1985, i, pp. 41-42, 105-6; Descartes 1991, p. 344.

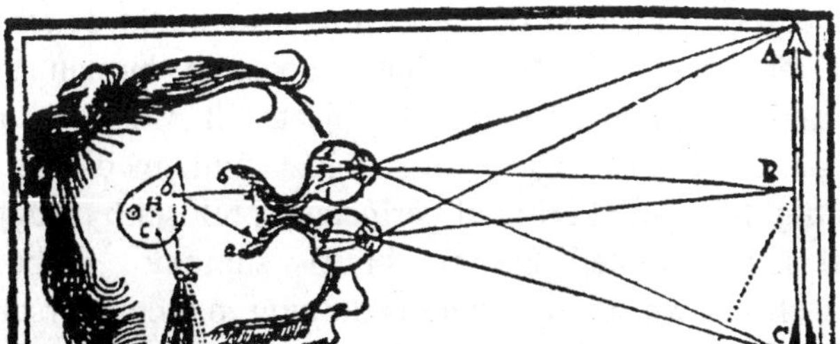

5.1 Descartes' illustration of the sensory perception of an arrow as coordinated by common sense (Descartes 1677, by permission of the Houghton Library, Harvard University)

representation of the object formed in the brain by the concerted operation of the sense of sight and common sense. The idea could thus become an object of intellectual contemplation.

Considered apart from the nature of the stimuli, the modern 'idea' and the ancient 'form' did not significantly differ from one another, as Gassendi noted in his *Institutio Logica.* These terms, he explained, had their common 'root in the phantasy or imaginative faculty'. Like other terms such as 'conception', 'preconception', 'anticipation' and 'phantasm', they were used by various authors to denote what 'is held in the mind' and 'is present to the mind, indeed, is thrust before it almost, when we think'.[10]

10 Gassendi 1981, pp. 83-4.

The age old doctrine of the soul as a recipient of external stimuli was thus reinforced by notions that associated human thought with simulation and imaging of external objects. It was one of the least disputable parts of the legacy of ancient and medieval philosophy. Gassendi, a highly reputed critic of peripatetic philosophy, suggested that it could be taken for granted. He thus briefly summarized the doctrine by noting 'the well-known saying "there is nothing in the understanding which was not first in the sense", and the dictum that the understanding or the mind is a blank tablet on which nothing has been engraved or painted'.[11] The common dictum was not apparently endorsed by every scholar. Admittedly some claimed 'that ideas have been imprinted in the mind by nature'; but it was nevertheless well established, for those who contested it 'are very far from proving what they say'.[12]

Gassendi might have addressed his critical note to Descartes and his disciples. On several occasions Descartes boasted to have deduced some principles of physics from a-priori metaphysical ideas.[13] In his *Meditations on First Philosophy*, he suggested, moreover, that the sensory faculties that human beings shared with other animals were entirely separated from the intellectual faculties that distinguished humans from other animals. He thus explained that 'the power of imagining which is in me, differing as it does from the power of understanding, is not necessary constituent of my own essence, that is, the essence of the Mind'.[14]

It appears, however, that Descartes did not intend his writings on the human soul to refute the received view that the input of the senses was a necessary component of human learning. In his *Rules for the Direction of Our Intelligence*, he ridiculed 'those Philosophers who take no account of experience and think that truth will spring from their brains, like Minerva from the head of Jupiter'.[15] For the mind to be aware of the existence of any other thing, including, arguably, a benevolent God, the power of understanding depended on the powers of imagination and sensation that characterized the person as a corporeal animal.[16]

---

11 Ibid., pp. 84-5.

12 Ibid., p. 85.

13 In a letter to Mersenne, May 27, 1638, see Descartes 1985, ii, p. 141

14 Descartes, 1985, ii, p. 51.

15 Ibid., i, p. 21.

16 Gaukroger 1994; Wickes 1994.

The vital role of the senses was often considered in light of the aims of philosophical learning. Gassendi arguably stressed more than Descartes that human knowledge originated in experience, but he did not assert the independent authority of the senses as knowledge-production faculties. Empirical information only provided the starting point of the philosopher's search for knowledge. In his *Institutio Logica*, he defined logic as the art of proper thinking. The aim of this art was the cultivation of four basic skills. Gassendi discussed these skills in accordance with the common assumption that philosophical research essentially involved intellectual abstraction that aimed to arrive at the definition of nature. The first skill pertained sensory perception, and consisted of forming adequate images or ideas of things. The other three skills presupposed the independent authority of the intellect, and consisted of forming adequate propositions, drawing correct inferences, and coherently ordering arguments in the course of inquiry and the course of instruction.[17]

Sensation and imagination provided the intellectual faculties with concrete images of things that formed the subject matter of philosophical discourse. Gassendi accordingly explained that the 'imagination' was 'able to encompass the whole description of a thing' without the intervention of the intellectual faculties, 'for example, of the sun, moon, clouds, thunder, earth, water, men, horses, plants, flowers, stones; in short, all things which we see, touch, smell, and so on, as they present and offer themselves to the senses'.[18]

By means of analysis, the various properties of the concrete image were grouped together or differentiated from one another, and thus gradually led by a series of analyses to prove a clear understanding of the nature underlying the concrete image. The student could conversely begin with a particular nature, and gradually compose a more concrete image of an instance of that nature.[19] As an illustration (Figure 5.2), he proposed a sequence from Socrates to Essence, taking note of 'the *differentiae* which have been set down on either side; for those which come first on the

17 A synoptic view of these skills is presented in the Preface, see Gassendi 1981, pp. 80-2.

18 Gassendi 1981, p. 85.

19 For detailed examinations of the twofold method, in reference to its exploratory, demonstrative and pedagogical functions, in the later Middle Ages and the Renaissance, see Randall 1940, 1961; Wallace 1959, pp. 36-43; Gilbert 1960; Crombie 1971, 1994, i, pp. 229-309; Hintikka and Remes 1974, Ch. 9.

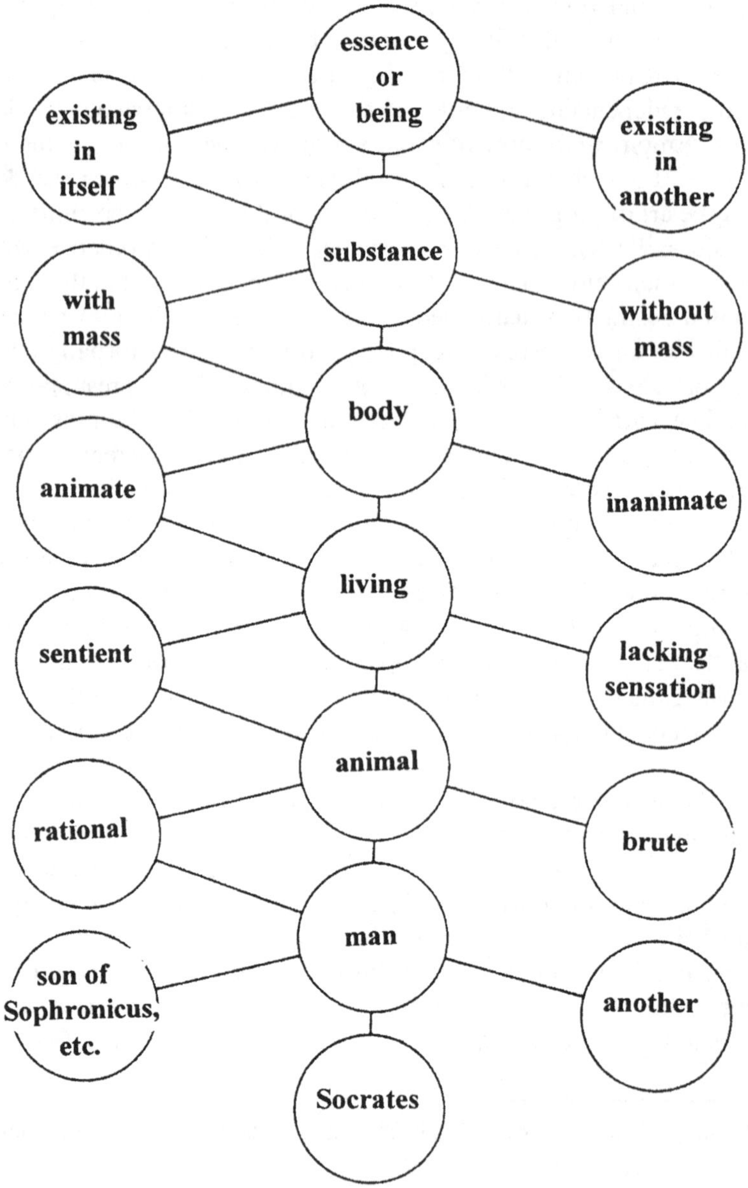

5.2 **A reconstruction of Gassendi's diagram of the methodical structure of analysis and synthesis,** with a translation from Latin (Gassendi 1981, by permission of the Houghton Library, Harvard University).

left hand side are those which take away from Socrates in order to arrive at Being'. The opposite direction of synthesis began with:

> Being and by adding these same *differentiae* arrive at Socrates. For Being, in and by itself, is Substance; Substance endowed with mass is Body; animate body is Living; a living thing endowed with feeling is Animal; a rational animal is Man; 'this' man, in this case the son of Sophroniscus, teacher of Plato and so on, is Socrates.[20]

Gassendi's simple pedagogical illustration thus disclosed the distinctively philosophical perspective on the world and the human capacity to understand it. The philosopher's world was comprised of many different 'beings', each of which was endowed with various properties and a specific nature. The distinction between the properties of a being and its nature corresponded to the distinction between the human senses and the human intellect, respectively. The senses were responsible for generating an image of the properties of a particular being. The intellect was responsible for defining its nature, and was therefore involved neither in determining the properties of a being, nor in determining how they were related to one another. Rather, its aim was to determine how the properties of a particular being were related to its particular nature.

This philosophical model of the division of labor between the senses and the intellect does not fit in with the modern distinction between empiricism and rationalism. The model was embraced , for example, by the alleged empiricist Hobbes and the alleged rationalist Descartes. Hobbes noted that the senses produced coherent images of highly varied complexity. 'For example, a man that is present in a foreign *city*, seeth not only *whole* streets, but can also distinguish particular *houses*, and *parts* of houses'. These sensory images could later be blurred by memory and imagination; yet 'at the time they are produced by sense' they retained their structural 'coherence or consequence'.[21]

According to Hobbes, concrete images of physical objects were generated by the senses without intervention from intellect. Thus, they were typically 'required in a Witnesse'. The witness' report was based on 'nothing else, but Sense and Memory,...as when we see a Fact doing, or

---

20 Gassendi 1981, pp. 88-9.

21 Hobbes 1840, iv, pp. 12-13, 15.

remember it done'.[22] The witness' knowledge of facts involved neither intellectual abstraction nor reasoning, and was therefore 'absolute'. The 'knowledge required in a philosopher', in contrast, was 'conditional', since it was based on presuppositions and inferences. The rational employment of the intellectual faculties enabled the philosopher to comprehend facts which the witness could merely record.

Hobbes illustrated the distinction using an example from geometry. A witness could identify a particular figure as a circle, but he or she would not understand its properties. A philosopher would know the geometrical nature of the figure and could therefore conclude that 'any straight line through the Centre shall divide it into equall part'.[23] Human understanding thus began with what was 'known to us' and culminated in grasping what was known 'to nature'.[24]

The immediacy of sensory perception, its authenticity, and its critical role in delivering to the intellect a well defined concrete object of contemplation, were commonly associated with the initial stage of philosophical research. Referring to his study of magnetic phenomena, Descartes recommended that the researcher initially 'carefully gathers together all the available observations concerning the stone in question, then he tries to deduce from this what sort of mixture of simple natures is necessary to produce all the effects which the magnet is found to have'.[25] In light of the critical role of the senses in setting the groundwork for philosophical reasoning, Descartes advised that

> it is better to resort only to those [observations] which, presenting themselves spontaneously to our senses, *cannot be unknown* to us if we reflect even a little. The reason for this is that the more unusual observations are apt to mislead us when we do not yet know the causes of the more common ones, and the factors on which they depend are almost always so special and so minute that it is very difficult to discern them.[26]

As shown in the previous chapters, philosophers regularly contested the authority of specific explanations proposed by their

---

22 Hobbes 1975, p. 117.

23 Ibid., p. 117.

24 Quoted from Madden 1960, p. 109.

25 Descartes 1985, i, p. 49.

26 Descartes 1985, i, p. 143; my emphasis. See also Garber 1993, 2001, pp. 85-110.

predecessors and colleagues. Disagreements were nevertheless well regulated by a widely shared notion of the aim of philosophy and by a widely shared image of mental operations that enabled philosophers to achieve their common aim. Descartes, Gassendi, and Hobbes were among the ardent critics of the scholastic tradition of the late Middle Ages. Their common adherence to scholastic views of the basic division of labor between the senses and the intellectual faculties demonstrates, however, how deeply these views were entrenched in the tradition of philosophical learning. During the 1670s, Newton indirectly challenged these views by demonstrating a new mode of explanation that was no longer predicated on the alleged capacity to ascertain the nature of things. At the same time, Locke addressed these views directly, and began to elaborate an alternative model of human understanding that would uproot and transform them.

Understanding the World in Which We Live

Newton gradually released his alternative approach to the philosophical study of nature during exchanges with his early critics. Locke immediately announced his intentions to propose an alternative in his introduction to the *Essay.* His work was designed to address the topics that were traditionally included in treatises on logic, method, and the human soul, such as cognition and the proper employment of the intellectual faculties, human learning, knowledge and its growth, and the quest for truth. His work intended to show, however, that it was neither necessary nor desirable to examine these topics in reference to the nature of things.

Philosophers were primarily concerned with the search for philosophical knowledge, rather than with the manner by which ordinary human agents gained knowledge about the world in which they lived. Locke's *Essay* aimed, in part, to reverse these priorities:

> Our business here is not to know all things, but those which concern our Conduct. If we can find out those Measures, whereby a rational Creature put in that State, which Man is in, in this World, may, and ought to govern his opinions, and Actions depending thereon, we need not be troubled, that some other things escape our Knowledge.[27]

[27] Locke 1975, i.i.6, p. 46.

In this context, Locke defined the human state as widely as possible, to include any aspect that appeared relevant to the 'great Concernments' of human agents, with respect to worldly 'conveniences', 'Information of Vertue', and 'the comfortable Provision for this Life', as well as 'the Way that leads to a better' hereafter.[28]

Locke repeatedly noted throughout the *Essay* that problem-solving activities in everyday life provided a sufficiently adequate frame of reference for a comprehensive study of human understanding and knowledge. He illustrated this point in his introduction, for example, in reference to the sailors' use of a depth-finder. "Tis of great use to the Sailor to know the length of his Line, though he cannot with it fathom all the depths of the Ocean'. This measuring instrument admittedly did not enable the sailor to ascertain the absolute truth; yet it was nevertheless of vital importance for the sailor in his endeavors 'to direct his Voyage, and caution him against running upon Shoals, that may ruin him'.[29] Problem-situations in everyday life were crucial experiments that time and again demonstrated that the search for practical solutions was simply more valuable than the quest for certainty. Elsewhere in his *Essay,* Locke accordingly noted that:

> Man would be at a great loss if he had nothing to direct him but what has the certainty of true knowledge. He that will not eat, till he has Demonstration that it will nourish him; he that will not stir till he infallibly knows the business he goes about will succeed, will have little else to do, but sit still and perish.[30]

Everyday life did not, however, merely demonstrate that ordinary agents preferred practical knowledge to the conclusions of philosophical inquiry. It suggested, in addition, that human agents who were philosophically illiterate systematically cultivated their cognitive and intellectual capacities. The sailor's use of a depth-finder, for example, involved searching for evidence, understanding this evidence, and reasoning about the structure of a tool and its applications.

It appeared, moreover, that the tasks routinely performed in adult life presupposed remarkable improvements of cognitive and intellectual

---

28 Ibid., i.i.5, p. 45.

29 Ibid., i.i.6, p. 46.

30 Ibid., iv.xiv.1, p. 652.

capacities during childhood. Locke thus noted, for example, that newly-born infants could barely control their sensory perception. Their sense-organs were physiologically functional, but their experience was indiscriminate – '*Light*, and *Colours*, are busie at hand every where, when the Eye is but open; *Sounds*, and some *tangible Qualities* fail not to solicite their proper Senses, and force an entrance to the Mind'. The ability to filter through the rich input of the senses and associate different colors and sounds with particular objects required methodical efforts. A child who was raised in a room with a painting of a landscape or a clock 'will have but a confused *Idea* of all the Parts they are made up of, till he *applies himself with attention,* to consider them each in particular'.[31]

Advancements in object recognition and manipulation were obviously the function of education; yet the latter was apparently not based on the assimilation of the traditional methods of philosophical learning. Older children often failed to distinguish between different objects simply because their education was very limited. For example, they routinely confused gold with brass because they superficially learned from their elders that gold was 'a thing with a bright shineing yellownesse'. A skilled craftsman or a chemist, by contrast, could easily discriminate between the two metals by testing their specific properties, such as 'weight', 'flexibility', 'fusibility', 'fixednesse', or 'aptnesse to be dissolved' in different solutions.[32] Pointing out the poverty of scholastic doctrines on 'substantial forms' in particular, Locke noted that any 'Smith, or a Jeweller commonly knows better than a Philosopher' the qualities 'observable in Iron, or a Diamond'.[33]

The emphasis on practical knowledge and its value in Locke's *Essay* reflected and addressed concerns of ameliorating civil life. This was routinely expressed in the writings of humanist scholars, particularly critics of the scholastic tradition of European universities throughout the 17th century. As previously shown, fascination with the prospects of technological progress led some scholars to reconsider the values that were traditionally associated with the academic culture of philosophical disputations. Notable initiatives were designed, however, to reform philosophical education – rather than renounce its traditional aims – and

31 Ibid., ii.i.6-7, pp. 106-7.

32 Locke 1990, p. 2; see also Locke 1975, iii.vii.50, p. 470.

33 Locke 1975, ii.xxiii.3, p. 296-7.

turn it into an effective instrument for the advancement of human prosperity.

Bacon's *New Organum* was perhaps the most celebrated attempt at demonstrating that genuine knowledge of the nature of things inherently empowered human industry. His new method, he proclaimed, enabled philosophers 'to establish and extend the power and dominion of the human race itself over the universe', and thereby to 'recover that right over nature which belongs to it by divine bequest'.[34] Descartes similarly professed in his *Discourse on the Method* that 'the general notions concerning physics' he had established revealed 'the possibility of gaining knowledge that would be very useful in life, and of discovering a practical philosophy to replace the speculative one taught in the Schools'. His new method could be applied to the study of 'the force and actions of fire, water, air, the stars, the heavens', and of the works of 'our artisans'. It provided the key with which philosophers could become 'as it were masters and possessors of Nature'.[35]

These philosophical visions were in part based on the age old assumption that the definition of the nature of things disclosed essential qualities in virtue of how they came into being. Philosophical research could thus render the make-up of physical entities transparent to, and in control of, the human mind.[36] Locke's *Essay* espoused neither new visions nor their presuppositions. Rather than subordinating human behavior and industry to a method that promised the knowledge of absolute truths, he claimed that philosophers ought to model their learning on the mental dispositions and skills embedded in prudent actions of everyday life.

Philosophers traditionally taught that proper learning was guided by rules of inference that related the concrete images of sensation and imagination with the abstract ideas that defined the nature of things. However, children and adults apparently did not learn by this way how to manage their daily affairs. More specifically, their concrete images of specific objects did not form the foundation of learning by abstraction. Rather, these images were continuously transformed as a result of

---

34 Bacon 1961, iv, pp. 27, 114-5, 294. For recent studies of Bacon's conception of philosophy as a master of nature and its sources, see Pérez-Ramos 1988; Gaukroger 2001.

35 Quoted from Ribe 1997, p. 43, which offers a detailed study of Descartes' vision and its sources.

36 Cuomo 2000.

education and training. Images that were more effective for one reason or another better served their beholders. Images that were not useful were discarded or modified. These were the products as well as the tools of human learning in everyday life. The mind of ordinary human agents was not set to abstract essential qualities from these concrete images. Rather, it was set to construct these images as usefully as it could. Using the philosophical nomenclature of 'Ideas', Locke's *Essay* purported to show how ordinary people made up their minds.

Following Descartes and Gassendi, Locke suggested that the various mental items that were either recorded or processed by mind could be classified as 'Ideas'. 'It being that Term', he pointed out in the first chapter of the *Essay*, 'which, I think, serves best to stand for whatever is the Object of the Understanding when a Man thinks, I have used it to express whatever is meant by *Phantasm*, *Notion*, *Species*, or whatever it is, which the Mind can be employ'd about in thinking'.[37] Locke reiterated the definition in the second book of the *Essay*, where, following Descartes, he stressed the distinction between ideas as mental properties and the material qualities of sensory stimuli: 'Whatever the Mind perceives in it self, or is the immediate object of Perception, Thought, or Understanding, that I call *Idea*; and the Power to produce any *Idea* in our mind, I call *Quality* of the Subject wherein that Power is'. For example, 'a Snow-ball having the Power to produce in us the *Ideas* of *White*, *Cold*, and *Round*, the Powers to produce those *Ideas* in us, as they are in the Snow-ball, I call *Qualities*; and as they are Sensations, or Perceptions, in our Understandings, I call them *Ideas*'.[38]

Current interpretations of Locke's use of the term 'Idea' oscillate between two seemingly conflicting definitions. Scholars such as Aaron, and more recently Ayers, have claimed that Locke was committed to an 'imagist' doctrine, according to which ideas were mental objects.[39] According to Yolton, however, Locke's 'idea' was more akin to the modern 'concept'. It meant to designate 'an act of awareness' or a disposition that 'also has a content, a psychic, cognitive feature distinct and different from the objects seen'.[40] The discrepancy between these accounts

---

37 Locke 1975, i.i.8, p. 47.

38 Ibid., ii.viii.8, p. 134.

39 Aaron 1955, pp. 44-51; Ayers 1991, pp. 38-51.

40 Yolton 1970, p. 88, 1975, p. 160, 1975a, p. 384. For more recent reviews and commentaries on the philosophical meaning of Locke's 'ideas' in relation to the

is partially explained away once Locke's usage is construed to denote the products of human understanding as well as its tools. 'Ideas' as products were images, as Ayers has suggested. 'Ideas' as tools were associated with mental acts, as Yolton has suggested.

Locke's consistent and systematic use of the term 'Ideas' is manifested in his innovative model of human understanding, and especially in his departure from Gassendi's exposition in his *Institutio Logica.* In accordance with the received distinction between the methods of analysis and composition, Gassendi distinguished between 'simple' ideas that were recorded by the senses and the imagination, and 'composite' ideas that conveyed propositions. He thus explained that 'We use the expression *simple* imagination', for example, 'of Man, Sun', because 'we imagine a thing purely and simply in an by itself', 'without either affirming or denying anything of it'.[41] 'Composite' ideas were employed, he explained, when 'we do not simply imagine something and look upon it unclothed, as it were, but form some judgment concerning it by making either an affirmation or denial'. The mind thus 'creates a composite idea from simple ones'.[42]

Locke accepted Gassendi's distinction, but interpreted it in a radically different way. He narrowed the category of simple ideas, and expanded the scope of complex ideas accordingly. The new distinction which he thereby drew made enabled him to demonstrate that the concrete images that philosophers traditionally considered as given by sense and memory were in fact intellectual constructs; the products and tools of learning. He could then show that further abstractions that aimed to reveal the nature of things were unnecessary.

Locke narrowed the category of simple ideas by specifying it to sensory perceptions that were traditionally attributed to discrete sensory stimuli.[43] A simple idea denoted a singular empirical property. It was therefore 'unmixed' or 'uncompounded', and 'contains in it nothing but *one uniform Appearance*, or Conception in the mind, and is not distinguishable into different *Ideas*'.[44] Simple ideas, such as '*Yellow, White,*

---

two aforementioned positions, see Mackie 1985; Ayers 1991, pp. 13-59; Chappell 1994; Lowe 1995, pp. 35-47; Wolterstorff 1996, pp. 14-18.

41 Gassendi 1981, pp. 80, 83.

42 Ibid., p. 102.

43 Locke 1975, ii.i.3, p. 105.

44 Ibid., ii.ii.1, p. 119.

Heat, Cold, Soft, Hard, Bitter, Sweet', conveyed discrete sensations. The idea of a man or an arrow, for instance, traditionally attributed to common sense and imagination, was not simple, by Locke's definition, since it was comprised of many simple ideas.

Locke added to the five external senses another faculty, which he termed 'reflection', by means of which the mind took 'notice' 'of its own Operations', such as '*Perception, Thinking, Doubting, Believing, Knowing, Willing*'. Although reflection was not 'Sense, as having nothing to do with external Objects', it was comparable to sensation in that it was confined to bringing specific states or operations into consciousness. Since reflection was confined to mental states or operations, Locke suggested that it 'might properly enough be call'd internal Sense'.[45] Simple ideas thus represented rudimentary qualities of things, such as a distinct color or shape, or a particular taste.

The mental items which Locke excluded from the category of simple ideas were now included in the category of complex ideas. The idea of a piece of lead, for example, which Gassendi classified as simple, was classified by Locke as complex. The latter, he explained, comprised relations between, or representations of, simple ideas. Simple ideas were the resources of which complex ideas were made. Locke thus noted that by relating 'the simple *Idea* of a certain dull whitish colour, with certain degree of Weight, Hardness, Ductility, and Fusibility', the complex '*Idea* of *Lead*' was formed.[46] The human mind, in general, 'has the Power to repeat, compare, and unite' simple ideas 'to an almost infinite Variety, and so can make at Pleasure new complex *Ideas*'.

In accordance with the everyday life perspective espoused in his *Essay*, Locke envisaged the transformation of the resources of simple ideas into the representations of complex ideas as the 'Dominion of Man, in this little World of his own Understanding'. The mind was an industrious unit, a laboratory devoted to enhancing human understanding. The inner setting of the mind's laboratory, with which every human agent was endowed, could be compared to:

> the great World of visible things; wherein his [man's] Power, however managed by Art and Skill, reaches no farther, than to compound and divide the Materials, that are made to his Hand; but can do nothing

---

45 Ibid., ii.i.4, p. 105.

46 Ibid., ii.xii.5-6, p. 165.

> towards the making the least Particle of new matter, or destroying one Atome of what is already in Being.[47]

The mind's laboratory received its staple input – 'its simple *Ideas*' – only from two sources, namely '*Sensation* and *Reflection*'.[48] With respect to this input, the 'Mind is wholly Passive'. A person could not feel the heat of a fire and simultaneously assert that it was cold. The industrious mind did not choose its resources. It could only transform what it already received. These resources of simple ideas thus comprised the 'Beginnings' of intellectual work, 'and as it were materials of Knowledge'.[49] The output was comprised of complex ideas produced by the 'power', 'art', and 'skill' of human understanding.

Locke's model of the mind turned the received philosophical account of analysis and composition upside down. Traditionally, the concrete image of a specific object was considered a record of 'sense and memory'. It presented an image with an underlying nature examined by philosophical analysis. The further the analysis advanced, the less concrete the image became. Analysis distanced the philosopher from the realm of interactions between human agents and other physical objects. In Locke's model, by contrast, analysis began with a particular mental construct, a 'complex *Idea*'. The analysis of complex ideas consisted of differentiating between their simple components. Since simple ideas denoted specific sensory perceptions – rather than an intellectual abstraction – the analysis of complex ideas could be checked against empirical evidence. In Locke's new model, analysis rendered human interactions with the world more informative.

The model suggested, accordingly, a different account of explanation. Human understanding was bounded by representations of empirical properties, or 'simple Ideas'. The only way to explain empirical evidence was by relating simple ideas to each other in a manner that rendered them more intelligible. Explanations differed from one another in accordance with the selection and correlation of simple ideas. Causal explanations, more particularly, consisted of representations that were made of simple or complex ideas of cause and effect. Locke thus explained that once human beings were sufficiently aware of acquiring simple ideas

---

47 Ibid., ii.ii.2, pp. 119-120.

48 Ibid., ii.xii.1, p. 163.

49 Ibid., ii.i.25, p. 118.

by sensation, they could not fail to notice that their sense organs were stimulated by environmental factors. Although they did not necessarily understand precisely what these factors were, they already acquired the ideas of cause and effect. By reflecting over their own intellectual endeavors, human agents similarly could not fail to notice the distinction between actions and their consequences. The very production of ideas 'which did not before exist, hath thereby in our Minds the relation of a Cause, and so is denominated by us'. 'That which produces any simple or complex *Idea*, we denote by the general Name *Cause*; and that which is produced, *Effect*'.[50]

It was therefore unnecessary to search for an explanation that defined a particular nature. A representation that explained how specific empirical properties were related was sufficiently adequate, for all practical purposes that could possibly concern 'a rational Creature...in that State, which Man is in, in this World'. The explanatory value of a particular proposition critically hinged, therefore, on how well it represented empirical evidence. Since definitions of particular natures did not represent the empirical properties of objects, they inevitably propounded speculative explanations.

Locke more specifically noted that an explanation that was predicated on a definition of a particular nature did not adequately take into account how empirical properties changed as a result of the changing relations between physical bodies. 'The changes that one body is apt to receive, and make in other bodies, upon a due application, exceeding far not only what we know, but what we are apt to imagine'.[51] Referring to his favorite example, Locke thus noted that if a piece of gold were isolated 'from the reach and influence of all other bodies, it will immediately lose all its Colour and Weight, and perhaps Malleableness too'.[52] Explanations that were predicated on definitions of particular natures were therefore too empirically arbitrary to be credible.

As shown in the previous chapter, Newton's rules of experimental philosophy abrogated the explanatory value of definitions of the nature of things by stipulating that explanations must relate phenomena. Locke's model imparted the same verdict for the same reasons. Newton claimed that the study of the physical world could be advanced so long as it was

---

50 Ibid., ii.xxvi.1, p. 324; ii.xxi.1-4, 72-3, pp. 233-6, 285-7.

51 Ibid., ii.xxxi.10, p. 382.

52 Ibid., iv.vi.11, p. 586.

conducted within the bounds of experimental philosophy. Locke's model also explained how it could thus be advanced. The value of complex ideas depended on the 'power', 'art', and 'skill' invested in their construction. It could be measured by determining how useful these ideas were when employed to form a judgment of matters of fact, to improve a technique of experimentation, or to explain empirical evidence. Considering the prospects of experimental philosophy, Locke foresaw that 'the dissolving of silver in *aqua fortis*, and Gold in *aqua Regia*, and not *vice versa* would be then, perhaps, no more difficult to know, than it is to a Smith to understand, why the turning of one Key will open a Lock, and not the turning of another'.[53]

[53] Ibid., iv.iii.25, p. 556.

Chapter Six

Locke's Doctrine of the Faithful Mind

Introduction

The previous chapter has shown that Locke's *An Essay on Human Understanding* presented a radical alternative to the age old philosophical approach to the problem of human understanding. The alternative was remarkably similar to Newton's, though was independently conceived of in about the same period – the late 1660s and early 1670s. Locke and Newton both claimed that the traditional search for definitions of the nature of things was inherently unreliable, and they both proposed that the understanding of the physical world could and should be confined to representations that showed how empirical properties were related to one another.

Newton endeavored to show in his publications that his new approach derived its meaning and value primarily from the solutions it offered to specific problems in natural philosophy, such as the problem of spectral colors. Locke's new model of human understanding was not designed to be so specific. Broad studies, such as Locke's *Essay*, derive their meaning and value from ideals and images of humankind and its place in the world. The principal objectives of such studies are accordingly defined in reference to cultural expressions of basic moral values and sensibilities.

Locke's perspective on human life was thoroughly theocentric. Dunn suggested that the *Essay* 'was principally directed' toward the 'project of *taking* responsibility for the content of one's belief'. Given this objective, Locke's 'conceptions of moral agency and the scope and limits of human understanding were closely linked'.[1] Ironically, modern empiricist theories of knowledge have argued for the vindication of a dichotomy between scientific knowledge and moral judgments. The empiricist rationale, in a nutshell, has been that knowledge is determined

1 Dunn 1969, 1986, p. 53, 1992, p. 61.

by empirical evidence, and is therefore indifferent to moral values and objectives.

Historical studies have shown that Locke's moral and political views were based to an important extent on natural law, which was commonly accepted by theologians and philosophers as the foundation of civil law and morality. However, since Locke's studies of human understanding were construed in the context of modern empiricist theories of knowledge, historians have concluded that Locke – perhaps inevitably – failed to coherently relate his theory of knowledge with his commitment to natural law.[2]

Following the examination of Locke's new model as described in the previous chapter, this chapter examines the theological and moral considerations that guided Locke in developing his new model. The first section shows that Locke conceived of his new model in response to doctrines that, in his view, threatened the social institution of natural law. The principal threat presented by these doctrines was that natural law was constitutive of human nature and therefore innate. The derivation of law from customs presented another threat. The most intransigent threat was conveyed by the philosophical assumption that explication of law must follow the dictates of reason.

Locke's critique of the philosophical perspective on natural law in particular led him to examine the distinction between knowledge and faith and its implications with respect to the problem of human understanding. The theory of knowledge which he then developed on the basis of his new model implied that representations of the physical world were articles of faith rather than knowledge. The final section of this chapter examines how faithful these representations were.

The False Authorities on Natural Law

Locke's entire intellectual career evolved around practical and theoretical issues concerning the relationship between social order and public good on the one hand, and the rule of faith and public transactions of religious

---

2 See, for example, the conclusions of Laslett and von Leyden, in Locke 1965, pp. 81-2 and Locke 1958, pp. 60-72, respectively; Harris 1994, pp. 266-271; Wolterstroff 1996, pp. 138-142.

beliefs on the other.[3] Throughout the 17th century, these issues were debated in the context of political crises stirred by opposing coalitions of a theologically divided clergy and a socially and politically divided lay public. In the late 1650s, the breakdown of Cromwell's government of the Protectorate indicated that the Church of England, in practice and by legal force, no longer took full responsibility for the spiritual life of the English people as a whole.[4]

Locke was then a young fellow of Christ College, Oxford, and composed his first scholarly work, *Two Tracts Upon Government.* In this manuscript, Locke adopted the common view among Anglican clerics and statesmen who considered the sectarian doctrines of radical non-conformists as the most dangerous threat to the precarious regime of the commonwealth in the aftermath of the Civil War. Locke defended at length the absolute and arbitrary authority of the magistrate, and called for enforcing comprehensive conformity to the Church of England.

He recognized, however, that the solution to the problem of moral and civil order directly pertained to the governing of individuals' conscience, and noted that the legislative power of the magistrate, whether civil or ecclesiastic, could not effectively transcend the bounds of 'indifferent matters': those which were not commanded by God either through natural law or by revelation. It appeared, therefore, that a stable political regime hinged to an important extent on a solution to the problem of knowledge of divine rule.[5]

In the early 1660s, Locke served several offices at Christ College, and was appointed censor of moral philosophy in 1664. Following his *Two Tracts*, he set out to critically examine authoritative sources for the explication of natural law. In accordance with the common academic style

---

3 For the definitive account of the development of Locke's social and political thought, see Marshall 1994.

4 According to Cameron, the breakdown of the coalitions between the reformed clergy and their lay supporters was the primary factor shaping the course of the Reformation and its immediate consequences during the late 16th century and the early part of the 17th century throughout Protestant Europe (in Cameron 1992). For the state of affairs in England, see also Stone 1980; Spurr 1991; Sparkes 1991; Sommerville 1992; Stocker 1992.

5 For Locke's early political views in relation to the problem of conscientious conduct, see Dunn, 1969, pp. 11-18; Marshall, 1994, pp. 12-32; Harris, 1994, pp. 65-77.

of disputation, Locke discussed the problem in a series of short essays which were organized in relation to specific adversary claims whose cogency he aimed to question.

The teachings of the church on natural law were traditionally based on the stipulation that human agents were by nature endowed with conscience and reason.[6] Locke construed of the notion of innate conscience as a potential threat of anarchy, since it implied that individuals could legitimately regard their own conscience as a sovereign interpreter of the law. Denying this implication, Locke argued that moral habits should have been much more uniform had they all originated from universal moral precepts 'inscribed' in the 'heart' or 'mind'.[7] For a similar reason, he rejected the assertion that the law of nature could be derived from 'tradition', 'the consensus of mankind', or its 'the natural inclination'.[8]

Another doctrine claimed that the law of nature could be ascertained by reason, the 'faculty of the intellect by which it articulates discourse and deduces arguments'.[9] Locke disputed the doctrine on the basis of moral and theological considerations. He thus claimed that human reason could not be regarded as 'the maker of this law…unless we are willing to diminish the dignity of the supreme lawgiver'. Human reasoning in itself could not possibly reveal God's intention. It 'investigates' the evidence of the law, rather than dictates it, 'since it is only a faculty of the mind and a part of us'.[10]

Locke thus identified three erroneous doctrines of the authoritative source of natural law: Innate conscience, customs, and reason. In his final essay he proposed an alternative source of authority: 'experience'. He emphasized, however, that he did not suggest that the law could simply be identified by the senses. The law obviously did not manifest 'itself somewhere so palpably that we could either read it with our eyes, or explore it with our hands, or hear it proclaiming itself'.[11] Locke, however, neither clarified how experience differed from the senses in this context,

---

6 Locke 1999, pp. 11-19; J. Yolton 1956; Marshall 1994, pp. 30-2; Atherton 1998.

7 Locke 1999, pp. 101, 119, 139-151.

8 Ibid., pp. 169, 173-201, 235-237.

9 Ibid., p. 99.

10 Ibid., p. 101.

11 Ibid., p. 131. The religious significance of experience in the writings of experimental philosophers is also examined in Dear 1990.

nor explained how experience could be brought about to demonstrate the law.

Several years later, Locke composed a manuscript now known as Draft A of his *Essay*. In this manuscript, Locke began to develop a doctrine of human understanding confined to relations between 'simple ideas'. He briefly noted that the law of nature could be 'easily & clearly knowne as any other Relation'. The law of nature was ordained by God, but so was the entire universe. Understanding the divine law did not differ, in this respect, from understanding other things that God ordained.[12] A study of human understanding in general could therefore show how the law of nature in particular could be known.[13] As will be shown in what follows, Locke further elaborated on his critique of the erroneous doctrines and on his alternative approach to the problem of natural law in Draft B of his *Essay*, and finally in the *Essay* itself.

The Attack on Innate Ideas

Locke expanded upon his arguments against innate ideas and principles in the first part of Draft B, which he completed shortly after Draft A. The revised version of this part eventually became the first book of his published *Essay*. As Yolton convincingly demonstrated, the notion of innate moral precepts was widespread among English clerics throughout the 17th century. It was often invoked to support religious convictions that affirmed prevailing values of pious conduct, especially in connection with the dictates of conscience and natural law.[14]

The doctrine of innate ideas may seem to have contradicted the common philosophical dictum that 'there is nothing in the understanding which was not first in the sense', but it did not. The former doctrine was used to justify religious obedience, and was endorsed hand in hand with the received philosophical view of the sources of human cognition. Faculty psychology explained how the senses enabled human beings to acquire knowledge about the world, but did not purport to explain religious obedience. People learned about the world from experience, but were moral agents in virtue of their innate knowledge of divine law.

---

12 Locke 1990, p. 41-2. See also Locke 1975, iv.xvi.14, p. 667.

13 For the relation of Locke's manuscript on natural law to the early drafts of the *Essay*, see von Leyden, 'Introduction', in Locke 1958, pp. 60-66, 74-80.

14 Yolton 1956, pp. 28-48.

The two doctrines not only failed to contradict one another, but were considered to have commonly vindicated the distinction between knowledge and faith. Bacon, for example, pointed out that 'the light of nature' operated in two mutually exclusive ways: 'the one, that which springeth from reason, sense, induction, argument, according to the laws of heaven and earth; the other, that which is imprinted upon the spirit of man by an inward instinct, according to the law of conscience'.[15] The mind, he professed, could direct itself in one of two primary ways, depending on its proper objective, 'for the face towards reason hath the print of truth, but the face towards action hath the print of good'.[16] He accordingly distinguished between the 'natural prudence' of the investigator on the one hand, and the 'moral prudence' of the law-abiding subject on the other.[17]

Locke rejected this division of labor. Living in accordance with God's rule and studying His creation were not separate callings. Rather, they were sanctioned by a single duty, which was at the same time a privilege. Natural law was an integral part of God's dominion, and so was the physical world and the ways by which it operated. The duty and privilege of recognizing the world as God's dominion entailed knowledge and obedience. This was one of Locke's main claims in his early essays on natural law.

In Draft B and later in his *Essay*, Locke undermined the division of labor by turning the received view that knowledge originated in experience against the received view of innate ideas. Since the former view was widely appraised as a well established philosophical theory of the nature of the human mind, it rendered his critique of innate ideas highly effective. Thus, in addition to historical arguments against innate morality which he had earlier suggested in his essays on natural law, he now proposed new arguments that pertained to human understanding. Locke thus explained that the doctrine of innate ideas could imply either that newly born human beings were conscious of their innate ideas, or that the ideas they entertained later on in their lives were originally innate.

The first option was obviously indefensible, since newly born children could not spell out definitive ideas or propositions. The second option was also highly implausible, he suggested, because it implied that children knew something which they could not possibly articulate. He thus

---

15 Bacon 1961, iii, p. 479.

16 Bacon 1980, pp. 116-17.

17 Ibid., pp. 97-9, 146-51

professed that children who were not conscious of certain ideas or propositions simply failed to have them. It was nonsensical 'to say, that there are Truths imprinted on the Soul, which it perceives or understands not; imprinting, if it signify any thing, being nothing else, but the making certain Truths to be perceived. For to imprint any thing on the Mind without the Mind's perceiving it, seems to me hardly intelligible'.[18] As Locke further noted in his examination of the faculty of memory, human beings were not aware of certain ideas or propositions once they were '*laying up*' in 'the Repository of the Memory'.[19] They could not, however, forget something they never thought or perceived. Learning did not involve the recollection of allegedly innate ideas. Quite to the contrary, it involved the purposeful acquisition of new ideas.

Undermining the Authority of Customs

Locke's critique of the authority of customs in his early essays on natural law echoed the findings of historical and anthropological studies of human societies, which demonstrated that customs varied widely from one society to another. His critique also reflected the growing awareness among scholars of the corruptive effects of habits and customs on human learning. One of the most celebrated accounts of these corruptions was Bacon's doctrine of 'Idols of the mind'.

Bacon claimed that idols of the mind stemmed in one way or another from the lack of a proper discipline that could effectively separate between sense and reason on the one hand, and naturally and socially ingrained wants and desires on the other. The human mind could in principle grasp the nature of things, but it could also easily be diverted from its path by numerous passions which imbued every aspect of life, and which could barely be recognized or controlled. The 'Idols' deceived, he explained,

> by a corrupt and ill-ordered predisposition of the mind, which as it were perverts and infects all the anticipations of the intellect. For the mind of man (dimmed and clouded as it is by the covering of the body), far from being a smooth, clear, and equal mirror (wherein the beams of things

---

18 Locke 1975, i.ii.5, pp. 49-50. See also Atherton 1983; Ayers 1991, i, pp. 267-8.

19 Locke 1975, ii.x.2, p. 150.

> reflect according to their true incidence), is rather like an enchanted mirror, full of superstition and imposture.

Bacon cautiously noted, moreover, that these deeply rooted defects could not be completely eradicated. The 'doctrine regarding Idols cannot be reduced to an art; all that can be done is to use a kind of thoughtful prudence against them'.[20]

Locke, to some extent, shared Bacon's view of the human character. As he gloomily wrote to a friend, 'men live upon trust and their knowledge is noething but opinion moulded up betweene custome and Interest, the two great Luminarys of the world, the only lights they walke by'.[21] As shown in the previous chapter, however, Locke did not espouse Bacon's vision of the philosophical mastery of nature. Unlike Bacon, he did not consider passion as an inherent impediment to learning. Lacking a passionate interest, 'Man, however furnished with the Faculties of Understanding and Will, would be a very idle unactive Creature, and pass his time in a lazy lethargick Dream'.[22] As will be elaborated later, Locke systematically incorporated passions into his new model of human understanding.

Locke's new model suggested, moreover, that habits of thought were very useful as long as they were not regarded as authoritative. Cognitive habits were formed when people routinely applied the same complex ideas to definitions of their problem-situations. They became detrimental to learning when individuals mistook complex ideas for reality. Locke noted, for example, that the simple ideas of a 'flat Circle variously shadow'd' could be confused with the complex idea of a globe, because 'the Judgment presently, by an habitual custom, alters the appearances into their Causes'.[23] Foreshadowing Hume's claim that inductive and causal inferences were regulated by habits of thought, Locke explained that such confusions, by a 'settled habit, in things whereof we have frequent experience', were 'performed so constantly, and so quick'.[24] It was therefore 'not so strange, that our Mind should change the *Idea* of its

---

20 Bacon 1961, i, p. 643; iv, p. 431.
21 Locke 1976, i, p. 123.
22 Locke 1975, ii.vii.3, p. 129.
23 Ibid., ii.ix.8, p. 145.
24 Ibid., ii.ix.9, p. 146.

Sensation, into that of its Judgment, and make one serve only to excite the other, without our taking notice of it'.[25]

The immediate and unreflective application of complex ideas was nevertheless often extremely convenient. It formed an integral part of the proficient use of tools of thought. Locke therefore did not seek to minimize the effects of habits on human conduct. As long as people did not maintain such habits for the sake of defending 'a System, Interest, or Party', 'humane Knowledge, under the present circumstances of our Beings and Constitutions may be carried much farther, than it hitherto has been'.[26]

The Censure of the Dictates of Reason

Of the various doctrines Locke examined and rejected in his early essays on natural law, the rationalist doctrine that claimed that law could be ascertained by reason was the most coherent and plausible. Since it was closely related to philosophical objectives of rational definitions of the nature of things, it was highly respected by theologians, jurisprudents, natural philosophers, and mathematicians.[27] Locke could easily show that other doctrines were either incongruent with historical and anthropological evidence, or self-contradictory. He could not do that with the rationalist doctrine. It was the only doctrine he opposed on moral grounds.

The rationalist doctrine derived its plausibility from two widely accepted suppositions. The first attributed the quality of supreme reason to God, and claimed that God created the universe and established His rule in virtue of that quality. God could not be erroneous, and, in virtue of His supreme benevolence, could not mislead His human creatures. Since the latter were by nature rational, reason appeared to be the natural means to ascertain laws of nature.

The second supposition claimed that rational arguments were perfectly unambiguous, or univocal, and were therefore understood by rational human beings and by God in precisely the same way. The two suppositions thus implied that reason offered the most reliable means to ascertain God's rule. Other means failed to honor divine reason, underestimated his creation of rational beings, and, their relative ambiguity led to misinterpretations and distortions of God's rule.

---

25 Ibid., ii.ix.10, p. 147.

26 Ibid., iv.iii.6, p. 540.

27 Funkenstein 1986; Oakley and Urdang 1991; Marion 1998.

Several historians suggested that Locke rejected the rationalist doctrine because he adhered to a voluntarist theology that attributed creation to God's will and power.[28] The two theological doctrines were not, however, mutually exclusive, since it was commonly assumed that God's creation typified all His qualities, including omniscience and omnipotence. Human reason was part of creation, and could therefore ascertain God's rule as an expression of God's wisdom and will.

As noted above, Locke's objections to the rationalist doctrine were moral rather than metaphysical, and specifically addressed the problem of obedience. Precisely because reason was univocal, the rationalist doctrine did not adequately capture or affirm the moral relationship between ruler and subject, nor between authority and submission. According to Locke, human agents were supposed to adhere to the law because it was God's, rather than because it was rational. The commandment lost its moral force when its content was conceived of by reason alone.

In hindsight, it appears that Locke predicated his moral considerations against the rationalist doctrine on his views concerning the relation between the law of nature and the rule of faith. He clarified these views in his *Reasonableness of Christianity*, which was published in 1695. Locke explained that although the law of nature was intended for all mankind, 'some parts of that truth lie too deep for our natural powers easily to reach, and make plain and visible to mankind, without some light from above to direct them'.[29] God's special assistance came in two major historical revelations, the first to Moses, which gave rise to the Jewish theocracy, and the second to Jesus, which gave rise to the religion of Christianity.

The historical record of the human search for law so far suggested, then, that the institution of the law originated in the charismatic leadership of individuals, and not in philosophical explications of moral issues. Christianity owed its existence, he noted, to Jesus and 'his apostles; a college made up, for the most part, of ignorant, but inspired fishermen'. The dictates of reason neither inspired action, nor enforced obedience. 'Not every writer of morals, or compiler of it from others...can thereby be

---

28 Byrne 1991; Oakley and Urdang 1991; Yolton 1991. For studies of the differences between rationalist and voluntarist doctrines in the works of English natural philosophers and especially advocates of the experimental philosophy, see Henry 1992; Wojcik 1997.

29 Locke 1999a, pp. 141-3, 147-8.

erected into a law-giver to mankind'. 'Mankind might hearken' to the teachings of one philosopher or another; yet they could 'reject it, as they pleased, or as it suited their interests, passions, principles or humours: they were under no obligation; the opinion of this or that philosopher, was of no authority'.[30]

These comments show how completely moral action, for Locke, was ingrained in religious experience. They also shed light on his early doctrine, in the essays on natural law, that the law could be ascertained by experience. There he stressed that law could not simply be observed or heard through the senses. He apparently drew an analogy between the discovery of the rule of faith in the scriptures, and its discovery in the physical world. As shown in Chapter Four, Newton made a similar analogy. They both sought to extricate the explication of God's rule from the philosophical discourse on the nature of things, and to associate it with faithful apprehension of the meaning of the scriptures. They both reckoned experience to be the key to faithful apprehension of the rule of God's dominion. Unaware of the other's work, they began to explore this possibility at about the same time – the mid 1660s; one in Oxford, and the other in Cambridge.

Locke began to develop his systematic critique of the rationalist doctrine in Draft B, and eventually presented it in the fourth book of his *Essay.* His critique was comprised of a synthesis of his new model of human understanding and the traditional distinction between knowledge and faith. The novelty of the critique stemmed from his new model, and for this reason he presented the distinction between knowledge and faith as conventionally as he could. It appears that he simply reiterated the definitions of Descartes and Gassendi.

Descartes explained in his *Rules for the Direction of Our Intelligence* that knowledge could be safely grounded 'without any fear of deception' only on 'intuition and inference'. Intuitive ideas were immediately formed by 'an unclouded and attentive mind'. For example, 'anybody can see by mental intuition that...a triangle is bounded by just three lines'. Knowledge by inference consisted in 'necessary conclusion from other things known with certainty'. For this reason, demonstrative knowledge, unlike intuition, was not always 'self-evident'.[31]

30 Ibid., pp. 150-1.

31 Descartes 1985, i, pp. 13-5.

Locke similarly delimited knowledge to two 'degrees' namely, 'Intuition and Demonstration'; 'whatever comes short of one of these, with what assurance soever embraced, is but Faith, or Opinion, but not knowledge'.[32] Intuitive knowledge was attained by the effortless recognition of the certainty of a proposition, when 'the Mind is at no pains of proving or examining, but perceives the Truth, as the Eye doth light, only by being directed toward it. Thus the Mind perceives, that *white* is not *black*, That a *circle* is not a *Triangle*, That *three* are more than *two*, and equal to *one* and *two*'.[33]

The second degree of knowledge was confined to logical inferences from intuitions – 'in every step Reason makes in demonstrative Knowledge, there is an intuitive knowledge'.[34] For this reason, demonstrations – especially in mathematics – involved intricate reasoning and required prior learning and investigation. Relevant intuitions, which 'the Mind perceives' 'but not immediately', must be identified and brought to relate to one another in a logical manner. The validity of a demonstration was accordingly less transparent than that of intuition, especially to those who lacked the skills required for the discovery of the intermediate intuitions.[35]

Gassendi distinguished demonstrative inferences from 'probable' inferences that followed contingent or conjectural premises. The former issued in 'science'; the latter in 'opinion', or provisional assent.[36] Locke similarly explained that opinion was formed when various ideas neither clearly 'agreed' nor 'disagreed' with one another. Hence 'the *difference between Probability* and *Certainty*, *Faith* and *Knowledge*, that in all the parts of Knowledge there is intuition; each immediate Idea, each step, has its visible and certain connexion: in belief not so. That which makes me believe, is something extraneous to the thing I believe; something not evidently joined on both sides to, and so not manifestly showing the Agreement or Disagreement of those *Ideas* that are under consideration'.[37]

Given this seemingly unambiguous distinction between knowledge and faith, the question then was whether human agents could gain

32 Locke 1975, iv.ii.14, pp. 536-537.

33 Ibid., p. 531.

34 Ibid., p. 533.

35 Ibid., p. 531-532.

36 Gassendi 1981, pp. 144-9.

37 Locke 1975, iv. xv.3, p. 655.

knowledge of God's creation, or whether they could only form beliefs about it. Applying his new model to the problem, Locke's answer stood squarely against the traditional philosophical view, which Descartes and Gassendi endorsed. As shown in the previous chapter, Descartes and Gassendi followed the teachings of scholastic philosophy and assumed that concrete images of physical objects directly originated in sensory perception, and therefore could be considered as intuitions, at least in those cases where the evidence was unambiguous. Such intuitions were the means by which demonstrative knowledge about the physical world could be obtained. Gassendi pointed out as an example that the appearance of sweat on the skin demonstrated that the skin was porous. The intuitive perception led directly to demonstrative knowledge.

In contrast, Locke's model described intuitive sensory perceptions as confined to discrete empirical properties unrelated to one another, i.e., simple ideas. The concrete images of physical objects were representations of simple ideas. They could seem intuitive simply because they were ingrained in habitual judgments; yet they were nonetheless the products of the industrious mind. Since reason could not dictate how simple ideas were related to one another, representations of objects in the physical world were therefore more or less probable opinions or articles of faith, rather than items of demonstrative knowledge.

Predicated on his new model of human understanding and the conventional distinction between knowledge and faith, Locke's theory of knowledge thus implied that human beings could not gain knowledge of the world. Locke, however, was not a sceptic. As shown in the previous chapter, he was confident that the growth of experimental philosophy would continuously advance human understanding of the physical world. Its achievements had already set 'Monuments to the Admiration of Posterity'. Its progress nevertheless advanced human faith rather than human knowledge. Moreover, its contributions were likely to present robust representations of physical objects that were farther remote from the intuitive-like perceptions of ordinary people in everyday life, just as the expertise of a goldsmith or a chemist was manifested in 'Ideas' that gradually departed from the common sense of laymen.

Locke further clarified implications of his theory of knowledge by considering an example from mathematics. He explained that studies in mathematics did not purport to determine what happened in the world. They were rather deliberately confined to the discovery of intermediate ideas that led to the proof of a relationship between relatively abstract ideas. 'The Mathematician considers the Truth and Properties belonging to a Rectangle, or Circle, only as they are in *Ideas* in his own Mind. For 'tis possible he never found either of them existing mathematically, i.e. precisely true, in his life'.[38]

In general, the search for intuitive and demonstrative propositions involved the construction of complex ideas that were not 'intended to be the Copies of any thing, nor referred to the existence of any thing'. These complex ideas were composed of simple ideas that were related to one another by a logical connection that was 'not designed to represent any thing but it self'. For this reason, such a logically composite propositional construct was univocal, and could 'never be capable of a wrong representation, nor misled us from the true apprehension of anything, by its dislikeness to it'.[39]

Locke readily acknowledged that logical and mathematical certainties were applicable to representations of objects in the world. In his comments on education, he stressed that mathematical training was extremely useful both because it could be applied to many different areas of learning, and because it accustomed the student to rigorous thinking.[40] The application of mathematical knowledge to the study of the empirical world, however, presupposed a more or less probable representation of particular empirical properties. Neither logic nor mathematics could dictate how these properties were related to one another.

Intuitions and demonstrations did not in themselves render the world intelligible. The assumption that the law of nature, or for that matter, any other aspect of God's creation, could be dictated by reason – which had troubled Locke in his essays on natural law – appeared in his *Essay* to be based on a misconceived and misleading notion of the scope of intuitions. By classifying phenomena as intuitions, philosophers confused logic, mathematics, and other domains of demonstrative knowledge on the one hand, with the endeavor to understand how the empirical properties of objects in the world were related on the other.

Locke further noted the difficulty in demonstrating propositions that included ideas which could not be subjected to mathematical analysis. When experience provided the intellect with simple ideas of qualities that could not be quantified, 'we have not so nice and accurate a distinction of

38 Ibid., iv.iv.6, p. 565.

39 Ibid., iv.iv.5, p. 564.

40 Ibid., iv.iv.6, p. 565. See also Locke 1849, pp. 493-5, 1989a, pp. 234-7.

their differences, as to perceive, or find ways to measure their just Equality or the least Differences'.[41]

Conversely, propositions that included quantifiable magnitudes that could not be empirically measured were also unlikely to be demonstrated. The problem, Locke noted, was particularly relevant to studies of mechanical philosophy primarily concerned with 'the Size, Figure, Number, and Motion of minute Corpuscles'. Since these magnitudes 'cannot be observed by us in Particles of Matter, whereof each is too subtile to be perceived, it is impossible for us to have any exact Measures of the different degrees of these simple *Ideas*'.[42] The 'clearest and most enlarged understandings of thinking men find themselves puzzled and at a loss in every particle of matter'. Human knowledge 'is but a point, almost nothing in comparison of the rest'. The search for 'a perfect science of bodies' was a 'lost labour'.[43]

The Faithfulness of 'Ideas'

Coupled with his new model of human understanding, Locke's critique of the authority of innate ideas, customs, and the dictates of reason, restricted understanding of the empirical world in its entirety to complex ideas. As shown in the previous chapter, Locke claimed that of all the possible ideas that the human mind could construct, only those that represented relations between empirical properties were reliable as a means of understanding. Other representations, such as those proposed by mechanical philosophers, were logically probable but practically highly unreliable.

Modern commentators have interpreted Locke's notion of probability within the framework of an empiricist theory of knowledge, and identified it with the measure of empirical adequacy.[44] In this context, a particular belief was more probable than another if and only if it better fit the empirical evidence. More recently, Shapin has shown that Locke followed other scholars in approaching the problem of empirical adequacy

---

41 Ibid., iv.ii.11, p. 535, iv.iii.3, p. 539, iv.iii.19, p. 550.

42 Ibid., iv.ii.11, p. 535.

43 Locke 1975, iii.vi.9, 11, pp. 444, 445-6, iv.iii.22-24, 29, pp. 553-5, 560, iv.iv.14, pp. 570, iv.xvii.2, p. 669.

44 Osler 1970; B. Shapiro 1983, 41-3; Ayres 1991, i, pp. 104-24; Woolhouse 1971, pp. 16-24.

from a broader perspective that took into account the social qualities of the observer or witness.[45]

Values of empirical adequacy, however, are too narrow to determine the value of representations as a means of understanding physical objects. Empirical adequacy may pertain to the empirical properties that are included in a particular representation, but it does not apply to assessing the adequacy of the relationship between properties.[46] Moreover, values of empirical adequacy do not directly address the value of representation as an activity. As shown in the previous chapter, Locke explicitly considered complex ideas as products of the voluntary transformation of simple ideas. A representation is right or wrong not only in relation to empirical evidence, but also in relation to practical values and considerations that pertain to representations as products of voluntary actions. For example, when Locke considered in the introduction to his *Essay* the prudent behavior of a sailor as an illustration of the scope of learning and knowledge, he explicitly noted that the sailor's responsibility for the safety of the boat was the principal factor in assessing the adequacy of the sailor's measurements of the ocean's depth.[47] The empirical adequacy of a representation of empirical evidence cannot therefore provide sufficient reason for its acceptance or rejection.[48]

Modern empiricist reconstructions of Locke's notion of the adequacy of beliefs fail to elucidate how it included moral considerations. Moreover, they fail to address Locke's theocentric perspective on moral and practical aspects of the adequacy of beliefs. As shown in the previous section, considerations of practical theology were paramount in his entire

---

45 Shapin 1994, pp. 211-30.

46 In modern philosophy of science, this problem has often been discussed in relation to the inadequacy of the positivist model of deductive-nomological explanations. For example, it has been noted that a representation of the length of a pole that relates the length of its shade to the angle of the propagation of sunlight does not makes sense despite its empirical adequacy. Obviously, the length of the shade is a function of the length of the pole rather than the other way round (for philosophical discussions of the advantages and weaknesses of the positivist model, see Pitt 1988).

47 See the third section of Chapter Five.

48 This problem applies to any empiricist theory of knowledge that excludes moral values from the solution to the problem of theory-choice. For a critical examination of Popper's methodology of falsification from this perspective, see Ben-Chaim 1998a.

critique of the false authorities on natural law. For this reason, understanding the alternative proposed in his *Essay* hinges on clarifying his account of theological aspects of representations of the physical world. As noted above, Locke explicitly asserted that these representations were articles of faith rather than knowledge. How faithful, then, were they?

Early modern Christianity was comprised of different forms of faith in God. Common to all these forms of faith was the fundamental conviction that the Christian faith was no mere abstract knowledge of the truth, but an efficacious way of receiving God's gift. As Gilson pointed out, Christianity 'offered itself at one and the same time as a doctrine and as a practice, or, more precisely, a doctrine that at the same time carried with it the means by which it was to be put into practice'.[49] Within this general cultural framework, the faith of a Dominican scholar, for example, considerably differed from the implicit faith a medieval peasant. The faith of a Puritan townsman in England differed from the faith of a Catholic courtier in 17th century Italy.

The faithfulness of Locke's representations originated in part from Protestant, and especially Calvinist faiths in the scriptural word as the only authentic manifestation of the messianic message of Christ. It nevertheless opposed the faith that was associated with Puritan radicalism of the Civil War. It closely mirrored the faith of moderate Anglicans in its commitment to law and order. It appears, however, that it most distinctively reflected the humanistic values of the Christian laity that gradually came to dominate English society.[50]

The human mind as it was portrayed in Locke's *Essay* was passionate but disciplined, and refused to separate the search for truth from the satisfaction of worldly needs and desires. Its hedonistic elements were consistently tempered with a theocentric conception of the human agent as a rational animal.

In examining how faithful 'Ideas' were, Locke drew special attention to two types of sensations, namely '*pleasure*' and '*pain*'. These were further distinguishable by various 'degrees', such 'Satisfaction, Delight, Pleasure, Happiness, *etc* on the one side; or Uneasiness, Trouble, Pain Torment, Anguish, misery *etc.* on the other'. Unlike most other

49 Gilson 1991, pp. 28-29.

50 For studies on the impact of humanism in reformed English theology, see Morgan 1986; Todd 1987; and with respect to Locke, see Sparkes 1991; Marshall 1994.

'simple Ideas', pleasure and pain could only be perceived in association with other ideas. 'One or the other of them join themselves to almost all our *Ideas*, ... And there is scarce any affection of our Sense from without, any retired thought of our Mind within, which is not able to produce in us *pleasure* or *pain*'.[51]

Locke noted in his *Essay* that these sensations were both signs and effects of the rule of 'the infinite Wise Author of our being'. In his early essays on natural law, he explained that God 'who has made and built this whole universe and us mortals', governed the human 'faculties and powers', stirring 'the one with sorrow or delight, the other with pain and pleasure; He also can lift both together to a condition of the utmost happiness or thrust them down to a state of misery and torment'.[52] Human agents were gripped by God's rule even if they failed to comprehend it.

Philosophers envisaged the contemplation of truth as a renunciation of worldly sensibilities. Gassendi, for example, emphasized the sensual character of the mind, but nevertheless professed that the soul's quest for the pleasure of tranquility defined the ultimate end of contemplative life. Natural philosophy enabled the student 'to forget his own mortality, and not recognize that he himself is a man, that is a weak animal, and from the condition of his nature liable to innumerable evils and miseries'.[53] From Locke's theocentric perspective on human life, however, pleasure and pain appeared to be transmitters of God's rule, and therefore ceaselessly maneuvered the mind. If these sensations 'were wholly separated from all our outward Sensations, and inward Thoughts, we should have no reason to prefer one thought or action to another; Negligence, to Attention; or Motion, to Rest'.[54] Foreshadowing Hume's dictum concerning the priority of passion over reason, Locke noted that the sensation of 'Uneasiness' in particular was the principal motive of human action, including the 'Workmanship of the Understanding'.

By responding to the signs and effects of divine rule, human beings became aware of themselves as voluntary agents and began to take account of their environment. At the most rudimentary level of self-consciousness, human agents could not fail to recognize that the scope of sensations of pleasure and pain was too extensive and that their impact was too irregular

---

51 Locke 1975, ii.vii.2, pp. 128.

52 Locke 1958, p. 153.

53 Quoted from Sarasohn 1996, p. 66.

54 Locke 1975, ii.vii.3, p. 129.

to allow relatively stable behavioral patterns. If we could not reflect upon and discipline responses to these sensations, we would have been

> seldom at ease, and free enough from the sollicitation of our natural or adopted desires, but a constant succession of *uneasinesses* out of that stock, which natural wants, or acquired habits have heaped up, take the *will* in their turns; and no sooner is one action dispatch'd, which by such a determination of the *will* we are set upon, but another *uneasiness* is ready to set us on work.[55]

Rudimentary choices were involved in selective responses to specific sources of pleasure or pain. Yet choice presupposed discipline. With the formation of discipline, human agents began to accommodate themselves within environments governed by God's rule.

Locke considered discipline as a problem of the regulation of will. Volition, in general, consisted of the '*Power* to begin or forbear, continue or end several actions of our minds, and motions of our Bodies'.[56] There was a two-tier order to the function of will. The first-order function consisted of bringing about a response to a particular desire. The second-order function of will consisted of balancing the satisfaction of one particular desire against another. This involved rational choice, which Locke in turn considered the principal expression of human freedom.[57]

Those who declined the 'opportunity to examine, view, and judge of the good and evil of what' they were 'going to do', forfeited their liberty,

> for this is the hinge on which turns the liberty of intellectual beings...that they can suspend this prosecution in particular cases, till they have looked before them, and informed themselves whether that particular thing which is then proposed or desired lie in the way to their main end, and make a real part of that which is their greatest good.[58]

A notion of freedom that failed to account for disciplined choice and disregarded the regulation of desires was inconceivable; according to Locke: 'if to break loose from the conduct of Reason, and to want that

55 Ibid., ii.xxi.45, p. 262

56 Ibid., ii.xxi.5, p. 236; ii.xxi.15-16, pp. 240-241.

57 For a detailed study on the relationship between discipline and freedom in Locke's political works, see Mehta 1992.

58 Locke 1975, ii.xxi.53, pp. 348-349.

restraint of Examination and Judgement which keeps us from choosing or doing the worse, be *liberty*, true Liberty, mad Men and Fools are the only Freemen'.[59] At the opposite extreme of madness was the state of slavery, in which one person was forced to submit himself or herself to another person's will. The madman and the slave were both not fully engaged in determining their own will, and failed to reflect upon the adequacy of their responses. More particularly, their intellectual autonomy was impaired because they could not consider themselves as actively engaged in forming the ideas that informed their actions.

Considering rational choice as the distinguishing mark of human freedom, Locke presented an analogy to compare intellectual and physical actions:

> As it is in the motion of the body, so it is in the Thoughts of our Minds; where any one [i.e., idea] is such, that we have power to take it up, or lay it by, according to the preference of the Mind, there we are at liberty. A waking Man being under the necessity of having some *Ideas* constantly in his Mind, is not at *liberty* to think, or not to think...But whether he will remove his contemplation from one *Idea* to another, is many times in his choice; and then he is in respect of his *Ideas*, as much at *liberty*, as he is in respect of Bodies he rests on: He can at pleasure remove himself from one to another.[60]

Since ideas about the world were associated with specific value placed on pleasure and pain, the liberty to examine ideas was crucial to making the right choices.[61] In taking the liberty to examine ideas and form the right judgments, then, empirical adequacy was only one of many considerations that determined human will. More broadly, an agent's emotions engaged the mind, his or her will inspired the understanding, and, concomitantly, dispositions toward good and truth were brought to bear upon one another, while perception gained practical value and could thus guide behavior. Within this broader framework, which integrated thinking with acting in the world, Locke noted that 'judgement' and 'election' were interrelated by 'a bare action of the understanding, whereby a man, several objects being proposed to him, takes one of them to be best for him'.[62]

---

59 Ibid., ii.xxi.50, p. 265.

60 Ibid., ii.xxi.12, p. 239.

61 Ibid, ii.xxi.47, pp. 262-3.

62 Locke 1884, p. 299.

Ideas were the resources and tools with which human agents exercised their liberty to distinguish right from wrong and to act accordingly. In examining their problem-situations, human agents did not assume the position of disinterested spectators. Rather, the situations that their cognitive faculties identified were always already experienced and understood as environments of action. The world was God's dominion, and every perception, thought, or action adapted human agents – in one way or another – to God's rule. For this reason, Locke confidently proclaimed in the introduction to his *Essay* that 'Our Business here is not to know all things, but those which concern our conduct'.

Locke's critique of the authority of innate ideas and the dictates of reason, and his proposed model of the 'Workmanship of the Understanding', abrogated the division between the philosophical study of natural phenomena and faithful conduct. The mind, he explained, was not divided into different compartments with special powers and tasks. 'The fault has been, that Faculties have been spoken of, and represented, as so many distinct Agents'. The 'power of Thinking' did not operate 'on the power of Chusing, nor the power of Chusing on the power of Thinking; no more than the power of Dancing operates on the power of Singing, or the power of Singing on the power of Dancing', he stressed, since it 'is the Man that does the Action, it is the Agent that has the power, or is able to do'.[63]

Whatever people chose to do with their lives, their behavior involved the elaboration of representations that at one and the same time related empirical properties of the environment, and related their environment to perceived needs or desires. Ideas were the means by which human agents searched for truth, but human deliberation was always concerned with the good. Ideas originated in the sensations of the powers of God's rule, and were then transformed into the tools with which free agents managed their affairs in God's dominion. The intellectual response to the rule was indelibly the function of the sensations it triggered. This was how faithful 'Ideas' were.

---

63 Locke 1975, ii.xxi.18-20, pp. 242-243.

Chapter Seven

Experimental Philosophy: The Gospel According to Boyle

Introduction

The previous chapters have shown that Locke and Newton considered the rise of experimental philosophy as a turning point in the philosophical study of the physical world. They both conceived of causality as a relation between observable properties, and claimed, accordingly, that causal relations could be properly represented by statements of fact. Both professed that the aim of experimental philosophy was the discovery of these facts. Moreover, they both espoused a theocentric view of causality as an expression of God's government. Experimental philosophy thus offered an alternative to peripatetic and mechanistic explications of the nature of things that also complemented the study of God's revelations in the scriptures.

The close affinity between their initiatives suggests, moreover, that they both followed in Robert Boyle's footsteps. During the early 1660s, Boyle was widely recognized as the leading exponent of experimental philosophy. He was one of the founding fellows of the Royal Society, and was highly regarded for his views on the congruence of experimental philosophy with the Christian faith. He wrote more on these issues than any other 17th century scholar. Newton owned twenty-two publications of Boyle's, and Locke's library included fifty-nine of his publications.[1] Boyle's work is likely to have inspired and guided Locke and Newton more than any other's in developing their common approach to the study of God's creation since the mid 1660s.

In accordance with the central dogma of science studies, historians have assumed that Boyle conceived of his experimental philosophy as a solution to the problem of knowledge. However, historical studies have

---

1 These estimates are based on Harrison and Laslett 1965, and Harrison 1978.

failed to identify the distinctive methodological solution that Boyle's experimental philosophy allegedly offered. It appears, from this perspective, that Boyle's thought was rather eclectic. It has been associated with the tradition of philosophical scepticism that was engendered by reactions against the religious dogmas of the Reformation and Counter-reformation movements. It has been shown, in addition, that Boyle followed Bacon's inductive and experimental methods as well as Descartes' deductive methodology.[2]

As Hunter pointed out, 'in the vast amount of time that he spent in his laboratory investigating chemical and other phenomena, Boyle's career resembled that of a modern research scientist to a far greater extent than was the case with most of his contemporaries'.[3] It is highly unlikely, therefore, that the methodological views that Boyle shared with other philosophers of the 17th century can adequately explain his apparently unique commitment to experimental research. Moreover, in as much as Boyle's career resembled those of modern researchers of natural science, methodologies of knowledge fail to adequately account for the differences between the empirical practices of 17th century philosophers (or from earlier periods) on the one hand, and those that came to characterize modern science on the other.

Of the historical studies on Boyle's experimental philosophy, a pivotal contribution has been made by Shapin and Schaffer in their *Leviathan and the Air-Pump.*[4] Advocating a sociological perspective on experimental research, they have demonstrated that Boyle's career initiated an institutional change in experimental research. Rather than considering experimental research as an aid to the philosophical search for a definition of the nature of things, Boyle endeavored to demonstrate that it formed an autonomous learning discipline that centered upon the discovery and public transaction of 'matters of fact'. Focusing on Boyle's pneumatic experiments, Shapin and Schaffer have thus stressed that Boyle was not

---

2 The canonical study of methodological scepticism in the early modern period is Popkin 1979 (see also Popkin 1988). For Boyle's methodological views in this context, see also van Leeuwen 1963; B. Shapiro 1974; Henry 1992. Boyle's methodological eclecticism is detailed in McMullin 1990; Hunter 1995; and especially Sargent 1995.

3 Hunter 1999, p. 263.

4 Shapin and Schaffer 1985.

> 'a vacuist' nor...'a plenist', and he mobilized powerful arguments against the mechanical and nonmechanical principles adduced by those who maintained that a vacuum was impossible. What he was endeavouring to create was a natural philosophical discourse in which such questions were inadmissible. The air-pump could not decide whether or not a 'metaphysical' vacuum existed. This was not a failing of the pump; instead, it was one of its *strengths*. Experimental practices were to rule out of court those problems that bred dispute and divisiveness among philosophers, and they were to substitute those questions that could generate matters of fact upon which philosophers might agree.[5]

As a social institution, experimental philosophy was comprised of ideals, rules, and patterns of conduct designed to regulate interactions within the community of experimentalists, and to dissociate its internal affairs from philosophical debates about the nature of things.

In accordance with the central dogma of science studies, Shapin and Schaffer considered Boyle's initiative as a solution to the problem of knowledge. As will be shown in what follows, his innovative approach to experimental research and its public transaction was an integral and critical component of his alternative to the mode of explanation that was traditionally espoused by natural philosophers. Boyle's experimental philosophy offered a novel solution to the problem of explanation, and his leadership – in this capacity – provided a model which Locke and Newton embraced and further consolidated.

The analysis in what follows aims in part to complement the study by Shapin and Schaffer. Thus, rather than addressing Boyle's experimental work, which has been thoroughly examined in *Leviathan and the Air-Pump*, this chapter focuses on the cultural values and ideals that were conveyed in Boyle's conception of experimental philosophy as a vocation. As will be shown throughout the chapter, humanist notions of religious worship provided Boyle with the cultural means to envision a new model of the philosopher's calling.[6]

The following section examines Boyle's early writings on the ethics of religious service. The next two sections discuss how he applied these ethics to the study of the physical world. The next section examines

---

5 Ibid., pp. 45-6.

6 For the function of non-scientific role models and model disciplines as frames of reference for scientific innovations, see Ben-David 1960; Jardine 1991.

Boyle's arguably unprecedented assault on the idolatry of the philosophical study of the nature of things. The final section shows how Boyle revised mechanical philosophy to accommodate it to his conception of true worship. His novel conception of matters of fact as correlations of empirical properties, and of experiment as the principal means of investigating factual correlations, followed directly from this revision.

In the Service of God's Dominion

Philosophy as a vocation was concerned with the provision of authoritative explanations. Boyle's views of his own vocation as a natural philosopher originated from his reflections on a more fundamental vocation, namely service to both God and fellow believers. Boyle began to examine this religious issue during the 1640s, when he was in his late teens, in what seems to have been personal studies on the calling of men of his age and social status.

These studies formed an integral part of his upbringing in the tradition of Christian humanism, which catered to the families of higher echelons of English society by Protestant scholars since the final quarter of the 16th century.[7] As Todd pointed out, Christian humanism offered an educational experience that 'was biblical in its apologetic, eclectic in its sources, mundane in its concerns but religious in its goals, practical in its methodology, and activist in its approach...The moral reconstruction of the social order was its ultimate objective – and its supreme attraction for Protestant reformers'.[8]

Born to the Earl of Cork, the young Boyle was expected to assume a position of civil leadership. His eldest brother, Richard, succeeded their father and became the Second Earl of Cork in 1643, as custom dictated. Roger Boyle, who was six years his senior, was already engaged in military and diplomatic services. The scholarly interests of the young Robert Boyle could have led to a career within the Anglican Church. Given his social position and the economic means it afforded him, however, Boyle did not need the patronage of the Church. Moreover, in a religiously divided

---

7 For studies on Boyle's humanistic education, see Harwood's introductory essays to Boyle 1991; Oster 1993, 1994; Principe 1994a, 1995.

8 Todd 1987, p. 22.

society, taking the oath entailed ecclesiastic loyalties that restricted a gentleman's free pursuit of the public good, as he noted later in his life.[9]

Boyle's early writings on the ethic of religious service reflected the yearning he shared with other members of his family and the English gentry for public peace and order that would secure their estates during the political turmoil of the Civil War.[10] These secular sensibilities converged with the theological ideal of moderation that was espoused by Anglican scholars since the early part of the 17th century, in their response to the growing tensions between the conservative legacy of the Henrician Reformation and militant Puritanism. Anglican moderation was predicated on the overriding interest in avoiding disputes on doctrinal issues that were not fundamental to salvation. It emphasized institutional unity at the expense of doctrinal unity, and advocated accordingly a religious literacy that centered upon the cultivation of prudent piety as a rival to formal methods of disputation and proof.[11]

Boyle's early writings expressed the values of moderate Anglicanism in an idiom of devotional piety that was at once highly spiritual, practical, and oriented to the public sphere. It centered upon God's benevolence as the principal progenitor of human agents' moral integrity, self-discipline, and commitment to public prosperity. The dialectical relation between spiritual devotion and the generation of wealth was the focus of *The Aretology or Ethicall Elements of Robert Boyle*. Completed in 1645, this was one of the longest and most comprehensive of his youth works.

In some respects, *The Aretology* conveyed a highly conventional overview of Christian morality. It was thoroughly theocentric in conceiving of God as the ultimate focal point of human motives, reasons, and actions. Its underlying principle proposed that the overriding perception of the good in every domain of life ought to be derived from, as well as affirm, trust in God as the fundamental source of human happiness. Boyle emphasized the indispensable function of non-intellectual faculties in moral deliberation. In accordance with the forms of literacy that were advocated by moderate

---

9 Boyle 1772, iv, p. 2.

10 Jacob 1977, pp. 43-82; Oster 1993, 1994; Shapin 1994, pp. 130-92.

11 For studies on the affinity between Boyle's theology and moderate Anglicanism, see van Leeuwen 1963; Shapiro 1974, 1983; Greene 1981; Heyd 1981, 1988; Shapin and Schaffer 1985, pp. 283-331; Hunter 1989; Henry 1992; Mulligan 1994; Wojcik 1997.

Anglicans, he noted that 'all the Definitions, Precepts, and Disputation in the World...cannot compell the Will and the Affections, to Chuse and enbrace' God's gift. Rational moralists, unlike pious believers, might have 'so much of Vertu in their Tongs that they have none of it in their Harts'.[12]

In accordance with the teachings of Christian asceticism, Boyle further stressed that the moral import of faith consisted in part of the formation of a discipline that enabled individuals to overcome temptations through self-denial. Passion was the great facilitator of misery and felicity, and its proper discipline originated in faith. The inner life of the soul centered upon the love of God, while action in the public sphere was directed towards the glorification of God. Embracing the teachings of humanist reformers, Boyle claimed that social callings delineated the proper institutional setting within which the disciplined soul embarked upon the search for happiness.[13]

The elementary callings found expression within the intimate unit of the family, while other callings were molded in the context of the more impersonal relations of civil society. The 'Apostles tels us that if we provide not for our Family, we are worse then Infidels'; and 'He is but an useless wastful Droane, and unworthy of the Benefits of Humane society, whose endeavors in som honest particular Calling, do not som way or other Cooperate (and contribute) to the Good of the Common-wealth'.[14] Boyle emphasized the harmony between ascetic devotion and secular industry, and claimed that their common pursuit should not be obstructed by prevailing social privileges.

> Every Person that is likely to have any vacant houres from serious Employments; shud lern some indifferent Skill, in Limming, Turning, Watchmaking, or Som Manuall Vocation or other. I know this will be spurn'd at by our Gallants as a Proposition fit to be made to blue Aprons then Skarlet Cloakes. But Sure it is not so much below a Gentleman to do something, as it is below both a Man & a Christian to be Idle.[15]

---

12 Boyle 1991, p. 54.

13 For the importance of the notion of calling in the writings of Christian humanism, see Todd 1987.

14 Boyle 1991, p. 85.

15 Ibid., p. 244.

Although Boyle refrained from discussing political aspects of the crisis that engulfed the Anglican Church since the outbreak of the Civil War in his *Aretology*, he pointed out the broad political implication of the ethic of secular asceticism. Greed, the predominant enemy of the devoted soul, found its expressions in the politics of 'Macchiavillians', those who were 'apt to discountenace Vertu', and glorified 'the Splendor of their greatness' at the expense of 'enlightening their Understandings'. Alluding to the fall of Charles' regime, Boyle challenged the 'admired Subtility' of their power politics, noting that many Machiavellians 'by their heaping Projects upon Projects have made and climed a Ladder, at the top of which, when they thouht to have seated upon a Throne, they have found themselves upon a Scaffold'.[16] Greed for political power corrupted the commonwealth by generating a culture of manipulative and dishonest relationships. Such a culture would flourish especially when tolerated by a polite society of gentlemen preoccupied with taking idle joy in their special privileges.

Boyle undoubtedly enjoyed the benefits of inherited wealth, but, as a matter of ethical principle, he claimed that the value of social privileges ought to be measured by their utility in augmenting public welfare. A person of noble birth and station was not exempted from the duty of diligence, 'there being no reason that because a man is Steward of a great family, he shud cast off all care of Administration'. On the contrary, his responsibility was greater precisely because 'most is given of him'. Rather than making 'vacation his only vocation', a gentleman ought to be a 'useful instrument of the Public Good'.[17] The ideal political regime thus appeared to be one in which power converged with industry, and both were intended to cultivate 'by Labor of Body, or Toyle of Mind, or both', the gifts with which individuals were endowed in virtue of their nature and their faith. Alluding to the ongoing political clashes between Royalist and Parliamentarian factions, Boyle noted that the 'Heroical Men' that could make the most of these gifts were 'more frequently observed' in

> Common-wealth then in Monarchys…, Partly because that in Republics the way to honor and preferment lys more open to desert, which is a quickning Spur and a great incitement to Noble Spirits; and Partly too,

16 Ibid., p. 67.

17 Ibid., pp. 88, 116, 243.

> because the lesser Inequality of Men's Conditions in Common-welths, renders these Heroic Spirits more conspicuous.[18]

The Study of God's Dominion

How did the young Boyle envisage his own calling in the society of 'Heroic Spirits' which he portrayed in his early essays? Boyle completed *The Aretology* a year after he returned from his grand tour of the continent, accompanied by his tutor Isaac Marcombes. It appears that at that time he did not have a clear view of his vocation as a young gentleman. In 1649, however, Boyle wrote to his sister, Lady Ranelagh, about a turning point in his studies on moral theology. Again preoccupied as he had been before with the humanistic ideal of the pious and industrious gentleman as a guardian of the public good, he was now engaged in novel endeavors to apply it to his own life: 'I will now presume to entertain you with those moral speculations, with which my chemical practices have entertained me' concerning 'the theological use of natural philosophy, endeavouring to make the contemplation of the creatures contributory to the instruction of the prince, and to the glory of the author of them'.[19] Research into the chemistry of the natural world now incarnated the 'heroic Spirit' of the young gentleman.

As Hunter meticulously demonstrated, during a short period of four years between his completion of *The Aretology* and this letter to his sister, Boyle embarked upon an extensive study of treatises on natural philosophy. He concomitantly expanded upon his new scholarly interest by carrying out experimental projects. Boyle's new interests were apparently engendered in part by his association with Benjamin Worsley, a Puritan, social reformer, and entrepreneur who sought to establish a business in the manufacture of saltpeter. Boyle met Worsley in 1646, and was immediately attracted to his industrial project. Through his contacts with Worsley, Boyle became acquainted with Samuel Hartlib and his Puritan activist colleagues, who developed utopian and utilitarian schemes for the advancement of human learning, especially in areas that were related to agriculture, medicine, and commerce. These new acquaintances brought the philosophical writings of Bacon and of some of the leading chemists in Europe to Boyle's attention.

---

18 Ibid., p. 132.

19 Boyle 1772, v, p. 238.

Seeking affiliation with the more established centers of academic scholarship, Boyle settled in Oxford in the early 1650s. He soon became acquainted with natural philosophers who, unlike the Hartlib circle, were predominantly Anglican and Royalist, a notable a group of Oxford experimental physiologists who evolved around William Harvey and included Thomas Willis, Henry Power, Christopher Wren, and Walter Charleton. During this period, Boyle studied the mechanical philosophy that was associated with the pioneering works of Descartes, Hobbes, and Gassendi. He expanded upon his chemical research, and in the late 1650s, he embarked upon one of his most ambitious and celebrated research projects on the qualities of air. Joining him were Hooke and a craftsman Ralph Greatorex, who assisted in establishing laboratory investigations with the 'pneumatic engine' of air-pump.[20]

From the perspective on science as ethically neutral knowledge, Boyle's new intellectual preoccupation was portrayed by Hunter as a 'transformation', a 'complete, unreversed change'. The young 'moralist' turned himself into a 'scientist' during a critical period between 1645 and 1648. His writings conveyed 'the record of a sudden conversion experience'.[21] The historical evidence, however, suggests that Boyle found experimental research valuable for precisely the same reasons he had earlier highlighted in his essays on moral theology. The affinity between his early moral reflections and his newly chosen career is especially evident by his programmatic writings in *Some Considerations Touching the Usefulness of Experimental Natural Philosophy*.

Based on several manuscripts Boyle composed circa 1650, his *Usefulness* was first published in London in 1663. It was the longest treatise in which Boyle set out to explain the vocation of the experimental philosopher. It was also one of his most popular works, and was printed in two editions that year. Boyle added another volume shortly after, and the two-volume edition was published in Oxford a year later.[22] Locke owned the first edition, a copy of which Newton could find in Barrow's library. Newton obtained the enlarged edition, and, according to Harrison, its many

---

20 Hunter 1995. For other studies on Boyle's initiation into natural philosophy and his early connections with the various groups of scholars in this area, see Maddison 1969; Webster 1965, 1975, pp. 57-61, 498-500; Principe 1994a; Newman 1996.

21 Hunter 1995, pp. 66-7, and 1999, p. 262.

22 Boyle 1999, 3, p. xix-xxvii.

dog-ears and other markings suggest that Newton frequently consulted this work.[23]

Boyle stated in *The Aretology* that faithful conduct was primarily the function of the proper discipline of two fundamental faculties, namely understanding and will. In his *Usefulness*, Boyle considered the pursuit of faithful life in the specific context of the study of the natural world. Boyle reaffirmed his conviction that the human soul was 'capable of wearing the Glorious Image of its Author, being endowed with two chief Faculties, the Understanding and the Will'. It therefore followed that the predominant virtues of the natural philosopher were, '*To know the Truth, and to do Good*'.[24] Boyle accordingly portrayed the glorious image of God which the study of nature intended to disclose as a mirror image of the ideal student. The physical universe conveyed God's wisdom and goodness. It also testified to the infinite industry with which God exercised these qualities. The natural world was thus comprised of the meeting place of the student's intellect, will, and industry, with the same but indefinitely greater qualities of God.

The endeavor to realize this image distinguished, in Boyle's view, a new vocation of experimental philosophy from the traditional practice of natural philosophy 'taught in most schools'. Teachings on natural philosophy were typically comprised of 'a Systeme of the Opinions of *Aristotle*, and some few other Writers'. It was not 'very difficult to be Learned' and become a natural philosopher by studying these opinions as well as those of 'a few of the more *Current* Authors'.[25] The experimental philosopher, by contrast, was called to the ministry of God. The priesthood of the scriptures redeemed humanity by adapting human language to God's instructions. The natural world posed a comparable challenge before the experimental philosopher,

> for we may say of the world, as St *Austin* did of the Sacraments, that it is *Verbum visible*. And certainly, God hath never so confin'd himself to instruct Men by words or types, as not to reserve himself the liberty of doing it by things: Witness his appointing the Rainbow to Preach his

---

[23] Laslett and Harrison 1965; Harrison 1978, p. 109; Feingold 1990, p. 345.

[24] Boyle 1999, iii, p. 199.

[25] Ibid., p. 199.

> Goodness to all Nations, and fortifie the Faith of Mankinde against the fear of a second Deluge.[26]

Alluding to the Protestant theology of biblical exegeses, Boyle claimed that natural occurrences were potentially 'Sermons', the 'Imortal Preachers' that 'give all Nations occasion to say of them, as the Assembly at Pentecost did of the Inspir'd Apostles, *We do hear them speak in our Tongues the wonderful Works of God*'.[27] The aim of the experimental philosopher was not to speculate about the causes underlying the phenomena, but rather to ascertain the meaning of empirical evidence as a sign of God's benevolent industry.[28]

The Calling of the Priestly Philosopher

In drawing attention to the religious significance of the study of nature, Boyle followed the conventional analogy contemporary philosophers drew between the study of nature and the study of scriptures. As Funkenstein pointed out, 'never before or after were science, philosophy, and theology seen as almost one and the same occupation'.[29] Unlike Boyle, however, natural philosophers stressed the importance of keeping the philosophical study of God's creation apart from the institutions of the Christian religion. In accordance with the age old distinction between knowledge and faith, Bacon, for example, warned that blurring the division between the philosophical study of God's creation on the one hand, and the obedience to the articles of faith on the other, could produce 'not only a fantastic philosophy but also an heretical religion'.[30]

Hobbes more explicitly stated that the methods and conclusions of science were incommensurable with religious worship. The latter consisted of obedience to authority, instead of reasoning, and was motivated by fear and ignorance of uncertainties people encountered throughout their

---

26 Ibid., p. 233. For an elucidation of the religious significance of experience in Boyle's experimental philosophy, see also Dear 1990.

27 Ibid.

28 For a detailed study of Boyle's approach to the interpretation of the scriptures in relation to his natural philosophy, see Sargent 1995, pp. 112-22.

29 Funkenstein 1986, p. 3.

30 Bacon 1961, iv, p. 66. See also Rabb 1969.

everyday lives.[31] Spinoza similarly proclaimed that 'between faith or theology, and philosophy, there is no connection, nor affinity...Philosophy has no end in view save truth; faith...looks for nothing but obedience and piety. Again, Philosophy is based on axioms which must be sought from nature alone; faith is based on history and language, and must be sought for only in Scripture and revelation'.[32]

Galileo reaffirmed the division of labor between priests and philosophers in his celebrated vindication of the autonomy of his profession during his clash with the administration of the Roman inquisition: 'the intention of the Holy Ghost is to teach us how one goes to heaven, not how heaven goes'. In his *Letter to the Grand Duchess Christina*, Galileo explained more specifically how philosophers could examine scriptural texts that addressed the physical world only in light of existing knowledge. The ministers of the church, by contrast, treated the scriptures as sacred means of religious instruction. The study of 'created things that are very remote from popular understanding' could not, in his view, be properly evaluated by principles that were 'pertinent to the primary purpose of the Holy Writ, that is, to the worship of God and the salvation of souls'.[33]

The rational pursuit of truth did not conflict with religious worship so long as the division of labor between them was maintained. Assuming that the truthful explication of the natural order was the prerogative of philosophy, Galileo suggested that when a particular reading of scriptures contradicted demonstrative knowledge, an alternative reading ought to be sought.[34] Pascal offered a different perspective on the tension between philosophical knowledge and the Christian faith. God loving wisdom was infinite in its scope and magnitude, and could not be adequately comprehended by the methods of philosophy. The philosophical study of God's creation could at best lead to deism, which undermined the Christian faith. Drawing practical implications from his conception of the dichotomous distinction between faith and knowledge, Pascal relinquished his career as a mathematician and philosopher shortly after his religious

---

31 Hobbes 1975, pp. 409-10, 628-9; Herbert 1989, pp. 116-21.

32 Spinoza 1925, iii, p. 179.

33 Finocchiaro 1989, p. 93.

34 Shea 1986; Feldhay 1995, p. 14; McMullin 1998; Rudavsky 2001.

conversion in 1654 at the age of thirty-one, when he joined a circle of Jansenist theologians and recluses at Port Royal.[35]

The calling of the experimental philosopher was predicated on renouncing this common division. The endeavor to understand the physical world formed an important aspect of the worship of God's dominion, rather than an autonomous profession with its own methods and aims. Its aim was to reveal the sanctity of the material world. Natural philosophers who confined their studies to the elaboration of systems of opinion forfeited this sacred responsibility. Referring to ancient Hebrew, Roman, and Apostolic sources, Boyle professed that in the remote past, philosophy had been widely recognized as a form of worship. 'Philosophers of almost all Religions have been, by the contemplation of the World, mov'd to consider it under the notion of a Temple'. This ancient lore was now revived by experimental philosophy. The new gospel declared that 'the World' was 'a Temple', and that 'Man sure must be the Preist, ordain'd (by being qualifi'd) to celebrate divine Service not only in it, but for it'.[36]

Boyle suggested, moreover, that God's worldly temple provided a more suitable frame of reference for worship than the scriptures because it more directly revealed God's industry. God's benevolent will was undoubtedly indicated in the scriptural narratives; yet the human mind could only grasp its immense value by way of interpretation. The worldly temple was an object of human experience. Its qualities could not be reduced to the qualities of a text, however sacred it was. It appeared to be a well-functioning kingdom, and the goods it afforded were tangible, and could directly sustain the lives of God's creatures.

In his *Usefulness*, Boyle spelled out some of the implications of the sanctity of the world for the calling of experimental philosophy by making reference to a metaphor of a clock. This mechanical device was often considered by his contemporaries as a man-made image of creation. Historical studies have drawn attention to the methodological significance of the metaphor of the clock. According to Laudan, Boyle followed the footsteps of Descartes and presented the metaphor to illustrate the probabilistic status of mechanistic explanations, in contradistinction to the relatively indisputable status of empirical evidence. The latter was

---

[35] Davidson 1983, pp. 96-108; Ashworth 1986, pp. 143-4; Jolley 1998, pp. 369-70. For Pascal's endeavors to bring religious values to bear on the philosophical study of nature, see M.L. Jones 2001.

[36] Boyle 1999, iii, p. 237-8. See also Fisch 1953.

metaphorically illustrated by the reading of the time as shown by the clock's hands. The former resembled the speculative conjectures about the inner mechanism of the clock.[37]

As shown in the previous chapters, Newton and Locke explicitly addressed this methodological distinction to attack the speculative character of philosophical explications of the nature of things. Both claimed that the explanations of the experimental philosopher were more reliable because they were conveyed with statements of fact. As will be shown in what follows, Boyle discussed the metaphor of the clock on four occasions in his *Usefulness*, where he systematically sought to present lessons that would foreshadow the methodological views of Newton and Locke from the metaphor.

Boyle initially addressed the image of 'the famous Clock at *Strasburg*' as an example of a 'Masterpiece of Art'. The pleasure and wonder it evoked resembled the experience of those who contemplated God's creation. Alluding to the recent achievements in physiological research, in particular to Harvey's discovery of the circulation of blood, Boyle noted that the glorious character of a work of design evinced by 'a skilful Dissection' of an organism was immensely greater than the impressive qualities of 'curious works of famous Artificers'. He suggested, moreover, that Aristotle's special attention to the nature of living organisms might have stemmed from a similar experience of wonder and admiration.[38] The industrious student of nature was called to follow in the artisan's or anatomist's footsteps by carefully inspecting the workings of a device instead of speculating about it solely on the basis of appearances.

Worship, medical dissection, and artisanship commonly involved learning by doing. Practical and pragmatic attitudes towards learning effaced the distinction between observation and speculation. They characterized students who construed the world in reference to their self-perception as agents who sought to take part in the production of tangible goods. Boyle compared the worldly temple to a palace that was owned by a lord who allowed his servants to dwell on his property. While the idle servants remained in one section of the palace, the more industrious explored it in an attempt to make the most of the gift they received from their benevolent lord. Their explorations, not unlike the anatomist's

---

37 Laudan 1966. See also Rogers 1972; Shapin and Schaffer 1985, p. 24; Sargent 1995, p. 210.

38 Boyle 1999, iii, pp. 201-2.

dissections, stemmed from their desire to understand and ameliorate their living conditions. At the same time, their debt to their lord increased as they learned to appreciate the value of his gift. Boyle concluded that 'the knowledge of the works of God proportions our admiration of them'. The 'further we contemplate them, the more Foot-steps and Impressions we discover of the Perfections of their Creator'.[39]

Boyle accordingly reconsidered the metaphor of the lock to highlight the work-ethic that characterized the experimental philosopher. He thus compared the person who was merely 'taken with the rich Enamel of the Case, and perhaps with some pretty Landskip that adorns the Dial-plate', to the industrious student who, 'besides that obvious Workmanship that first entertains the Eye, considers the exactness, and knows the use of every Wheel, takes notice of their proportion, contrivance, and adaptation altogether, and of the hidden Springs that move them all'.[40] The existence of God

> is indeed so legibly written on the Creatures, that...even a perfunctory Beholder, that makes it not his business, may perceive it. But that this God has manifested in these Creatures a Power, a Wisdom, and a Goodness worthy of himself, needs an attentive and diligent surveyor to discover.[41]

The doctrines of a spectator who speculated about the mechanism of the clock seemed, in this context, to be the epitome of disgraceful idleness.

The clock metaphor illustrated, moreover, Boyle's revision of the natural theology argument from design. Traditional formulations of the argument were structured by inferring from the rational qualities of creation to the rational qualities of the divine architect and creator. The credibility of this inference hinged on the notion that rationality was a univocal quality. As noted in the previous chapter, this notion was crucial to various rhetorical strategies that sought to set theology in rational foundations. Boyle, by contrast, construed the argument from design in accordance with his humanistic theology of dominion. The ethical aspect of his theology was partly conveyed by the social ideal of the 'heroic Spirit' he had earlier portrayed in *The Aretology*. In his *Usefulness*, Boyle emphasized the

---

39 Ibid., pp. 205-6.

40 Ibid., p. 235.

41 Ibid., p. 236.

harmonious qualities of God's dominion as a system of relations that reflected God's power and benevolent wisdom. A clock, by analogy, could not work well unless appropriately powered. The quality of its work hinges on the coordination of its parts.

Power and structure were therefore the two principal dimensions of worldly affairs. Physical bodies or states could neither come into being nor be sustained but for the power that was somehow invested in them. Each and every physical body or a state constituted of a configuration, or a system of relations between more elementary bodies or states. While the metaphor invoked a mechanical image of the physical universe, Boyle noted in this context that the functional characteristics of created objects was not necessarily mechanical. There were chemical and physiological 'Contrivances' which did not readily accord with the clock-work imagery of gear mechanism.[42]

Boyle's revision of the design argument seems to have shaped Locke's approach to addressing the problem of natural law. Locke apparently read Boyle's *Usefulness* soon after it was published, and shortly thereafter composed his essays on natural law. He was by then familiar with Boyle's views on religion and experimental philosophy, since they regularly met at Oxford during the early 1660s. They shared an interest in experimental medicine and chemistry, two sciences which Boyle appraised as valuable venues for demonstrating God's benevolent rule.

In accordance with Boyle's moral theology of dominion, Locke depicted the natural world as a repository of divine goods, which 'provided a certain profusion of goods for the use and convenience of men'. 'Good, rich veins of gold and silver lie hidden in the bowels of the earth, and moreover arms and hands and reason, the inventor of machines, are given to men, with which they can dig them out. Yet from this we do not conclude that all men are wealthy'.[43] The human capacity and right to enjoy what was 'concealed in the bowels of the earth' were themselves divinely ordained.[44] Locke shared with Boyle a commitment to a theocentric ethic of devotion to the public good, and claimed, accordingly, that God's benevolence should not be abused by self-interest, since 'the things provided have been bestowed in a definite way and in a predetermined

---

42 Ibid., pp. 250, 254. For Boyle's conception of the relation between mechanics and chemistry, see Clericuzio 1990.

43 Locke 1958, p. 135.

44 Ibid.

quantity'.[45] Locke followed Boyle in proposing his solution to the problem of natural law, namely that experience provided human agents with a means to comprehend their role in God's dominion.

Boyle reexamined the clock metaphor in drawing the methodological implications of his theocentric image of natural order. He highlighted the hypothetical status of philosophical explications of causal mechanisms, as Laudan noted, but claimed that the public transaction of these hypotheses hindered the advancement of learning. Philosophers aimed to discover 'the true and adequate Causes of Things'. The 'grand Argument by which they use to confirm the truth of their Explications, is, that either the *Phaenomenon* must be explicated after the manner by them specified, or else it cannot at all be explicated intelligibly'. In practice, however, they commonly acknowledged the conjectural status of explications, and accordingly 'assign not precisely the true, but possible Causes of the *Phenomenon* they endeavor to explain'.[46] The chronic discrepancy between the promise of necessary truth and the actual delivery of hypotheses indicated that philosophers could not be trusted with the faithful representation of God's rule.

As shown in Chapter Three, Boyle noted in his treatise on colors that the pretentious aim to reveal necessary truths led to 'sectarian' divisions between philosophical schools. In his 1672 article, Newton disapprovingly labeled the philosophers' aim as their 'universal Topic'. His critique echoed Boyle's disapproval of the philosophers' 'grand Argument' in his *Usefulness*. Newton read Boyle's treatise on colors by the mid 1660s, after the publication of his *Usefulness*. By 1672, Newton confidently allied himself with the statements of England's leading proponent of the experimental philosophy. In his *Usefulness*, Boyle professed that 'what we' – the experimental philosophers – 'are to enquire after, is, how Things have been, or are really produced, not whether or no the manner of their Production be such, as may most easily be understood by us'.[47] Newton could not have more clearly indicated his agreement with Boyle when he stated to his critics that all that he aimed to know was how colors were '*de facto*' produced, rather than what their nature could possibly be.

45 Ibid., p. 211. See also Marshall, 1994, pp. 157-204, 192-326.

46 Boyle 1999, iii, p. 256-7.

47 Ibid., p. 257.

The Revolt Against Idolatrous Notions of Nature

Boyle's portrayal of experimental philosophy as a form of worship implied that philosophical studies that were incongruent with its distinctive means and ends committed idolatry. The aim of experimental philosophy was the study of God's universal temple. However, philosophers traditionally focused their studies on the nature of things. The philosophical study of nature was not necessarily compatible with true worship.

Shortly after his *Usefulness* was published, Boyle began to compose an essay that set out to show how philosophical concepts of nature were idolatrous because they failed to adequately demonstrate that God's 'workmanship' was the true object of human understanding. A couple of decades later, Boyle prepared this essay for publication, and *A Free Enquiry into the Vulgarly Received Notions of Nature* appeared in 1686.

Boyle's *Free Enquiry* complemented his *Usefulness* in demonstrating the extent to which he conceived his experimental philosophy as a turning point in the history of western philosophy. In Boyle's view, practically the entire tradition of natural philosophical learning that originated in the writings of ancient Greek philosophers was philosophically and religiously erroneous. The philosophical error consisted of conceiving of the physical order as an effect of essential properties of physical entities. Since the latter were part of the former, they could not coherently be considered its cause. Although some qualities could participate in generating other qualities, none of them could be considered the fundamental object of a causal explanation. The traditional search for a definition of the nature of things was therefore predicated on confusion between causes and effects.

The mistaken identity of the object of philosophical inquiry involved a religious error because the true object of human understanding was God's government. The fundamental misconception of philosophers originated in pagan notions of nature that were propagated by ancient Greek philosophers, and especially Aristotle, who failed to recognize that the world was a system of the works of God. Referring to Aristotle's definition of nature as 'the principle of motion and change', Boyle pointed out 'that the definition of a thing, the most important in natural philosophy, should be clearly and accurately delivered'; yet Aristotle failed to meet this basic requirement because he conceived of the world as a system of natural things as well as a system of works of nature. Lacking a clear notion of a governor of the universe, Aristotle considered nature as cause and effect.[48]

The idolatry of nature was particularly apparent in the juxtaposition of the writings of ancient philosophers and prophets. The 'heathen' philosophers taught that nature, 'by her plastic power and skill, forms plants and animals out of the universal matter'; Moses, 'the divine historian', attributed 'the formation of them to God's immediate *fiat*'. The biblical history of creation 'has not a word of nature'.

> So likewise, though *Job*, *David*, and *Solomon*, and other *Israelitish* writers, do, on divers occasions, many times mention the corporeal works of God, yet they do not take notice of nature, which our philosophers would have his great vicegerent in what relates to them...And when St. *Paul* himself, who was no stranger to the Heathen learning, writing to the *Corinthians*, who were *Greeks*, speaks of the production of corn out of seed sown, he does not attribute the produced body to nature; but...adds that *God gives it such a body as he pleaseth, and to every seed its own body*.[49]

With respect to the problem of causality, then, the sources of monotheism stood squarely against the teachings of ancient Greek philosophers. In Boyle's view, Christian philosophers failed to acknowledge the dichotomy between these two positions. They adapted the teachings of the Bible to the language of pagan philosophy by combining nature with God – 'nature being God's vicegerent, her works are indeed his'. One of the aims of Boyle's *Free Enquiry* was to sort out the ensuing confusion by reinstating the hierarchical distinction between the natural world and its divine ruler. God 'is not only a self-existent and independent being, but the supreme and absolute Lord, and if I may so speak, the proprietor of the whole creation'. The subordination of nature to the commands of God was declared in the scriptures, which explicitly stated that the Lord was 'not only "the most high God", but " קנה *koneh*, possessing" (or, as our version has it, "possessor of") "heaven and earth"'.[50] The political economy of acts of appropriation indicated that the

---

48 Boyle 1772, v, p. 171.

49 Ibid., p. 172.

50 Ibid., p. 197. Boyle's quotation from Genesis 14:19-23.

properties of the physical world were the ownership of God rather than of any of His creatures.

Parting from philosophical discourse on natural causality, Boyle asked his philosophically learned audience to

> be excused, if with *Moses*, *Job*, and *David*, I call the creatures I admire in the visible world, the works of God, (not of nature) and praise rather him than her...; since among the *Israelites*, till they were overrun and corrupted by idolatrous nations, there was for many ages a deep silence of such a being, as we now call nature.[51]

Boyle realized that his critique of the idolatry of philosophical learning could be construed as an expression of a dogmatic prejudice against it. He therefore stressed that his work was itself the fruit of the 'free enquiry' of a scholar who sought to advance human learning, and who was well aware of the harmful effects of imprudent loyalty to the teachings of any particular church.[52]

Boyle's analysis of the problem of explanation in the context of the history of the relation between religious faith and philosophy demonstrates the scope of his impact on Newton. As shown in the previous section, Newton agreed with Boyle in addressing the corruptive effects of the public transactions of philosophical hypotheses on human learning. Newton implicitly affirmed the religious error of philosophical custom by choosing a sermonical style as the appropriate rhetorical mode of the presentation of his new theory on light and colors. He concomitantly initiated a series of studies on the corruptive impact of ancient philosophy on the teachings of the church.

In agreement with Boyle's views, Newton claimed that the scholastic tradition systematically misinterpreted God's works by elaborating metaphysical systems that undermined biblical notions of God as a benevolent ruler. In his drafts for the 'General Scholium' of *Principia*, Newton drew attention to the prevailing confusion of causal notions of nature with the true meaning of God as the supreme object of worship.

> He who shall demonstrate that there is a Perfect Being, and does not at the same time demonstrate that he is Lord of the Universe or Pantokrator, will

---

51 Ibid., p. 173.

52 Ibid., pp. 158-60.

> not yet have demonstrated that God exists. A Being eternal, infinite, allwise and most perfect without dominion is not God, but only Nature.[53]

In agreement with Locke's critique of the rationalist approach to the problem of natural law, Newton further explained that true worship entailed personal obligation to the sovereignty of the ruler on behalf of each and every human subject.

> We say, my God, your God, the God of *Isràel* the God of Gods, and Lord of Lords; but we do not say, my Eternal, your Eternal, the Eternal of *Israel* the Eternal of Gods; we do not say, my Infinite, or my Perfect: these are titles which have no respect to servants.[54]

Newton further explained in his early draft of the *Original Religions*, composed in the 1680s, that the ancient prophesies, which were originally written 'allegorically & morally with relation to piety & virtue', were later abused by 'men of corrupt minds' who, 'wresting the expressions of scripture to the opinions of philosophers, have brought the Christian religion many philosophical opinion to w[ch] the first Christians were strangers'. In his correspondence with Anglican scholar Thomas Burnet during the late 1680s, Newton claimed that Moses' account of the creation 'was not philosophical or feigned', but was later gradually corrupted by the interpretations of 'learned people' who deliberately set it 'above the understanding of meanest people'.[55]

In his *Free Enquiry*, Boyle examined the biblical sources on ancient origins of idolatry. The most ancient idolatry, he noted, originated in Egyptian cults, and 'seems to have been the worship of the celestial lights, especially the sun and the moon: that kind of *aboda zara*, עבדה צרה (as the Jewish writers call strange or false worships)'.[56] Newton reached a similar conclusion. According to David Gregory, who in 1694 discussed this subject with Newton, the latter held that 'Moses began a reformation but retained the indifferent elements of the Egyptians', 'who most of all

---

53 Hall and Boas Hall 1962, p. 363.

54 Newton 1999, pp. 940-1. For detailed studies on Newton's theology of dominion, see Force 1990, 1990a, 1994; Snobelen 2001.

55 Quoted from Iliffe 1994. See also Iliffe 1999; Markley 1999.

56 Boyle 1772, v, pp. 180-1.

debased religion with superstition and from them it spread to other people'.[57]

According to Boyle, the philosophical idols later developed by the Greeks included doctrines about animated entities that pervaded the physical universe. 'The great affinity between the *Soul of the world*, so much talked of among the heathen philosophers, and the thing, that men call nature', was conducive to the development of the philosophical concept of nature as the principle of motion and change. Moreover, the idolatrous assumption that 'this mundane soul was not barely a living, but a most intelligent and wisely active Being', later captured the religious imagination of Christian scholars.

Philosophical doctrines of an intelligent 'spirit' of nature, or a 'reasoning nature', were especially pernicious. They suggested a solution to the problem of God's rule by stipulating the existence of non-material entities that mediated between God and brute matter. The solution, however, only blurred the dichotomous distinction between ruler and subjects.[58] Boyle noted, as an example, that Galen, in his

> admirable treatise *De usu partium*, where he so excellently declares and celebrates the most wise author of things, was so far transported with the error which infected so many heathen philosophers, that he fancied the earth itself...had a certain soul or mind imparted to it by superior bodies,

as if they themselves were 'more pure', 'inhabited by a mind so much the better and more perfect than...terrestrial bodies'. Boyle accordingly stressed that neither the world, 'nor the soul of it – is the true God'.[59] Newton similarly professed in the General Scholium that God 'governs all things, not as the soul of the world, but as Lord over all...*Deity* is the dominion of God not over his own body, as those imagine who fancy God to be the soul of the world, but over servants'.[60]

---

57 Newton 1959, iii, p. 338.

58 Boyle 1772, v, pp. 183-185.

59 Ibid., pp. 181-2.

60 Newton 1999, pp. 940-1. However, in an essay Newton composed in the early 1670s, he associated light and the principles of vegetation and fermentation with a certain spirit, 'Natures universall agent, her secret fire, the materiall soule of all matter', the only conceivable cause of heat and motion'. Newton further noted that the 'Sun & Moon' were the surrogate 'father & mother' of the universal spirit, 'whence it inherits speedily ye first crown of perfection' (quoted from Dobbs 1991,

With the development of natural philosophy in late medieval universities, these idolatrous notions became 'vulgar' by forming an integral part of the common language scholars carelessly used. Boyle reprimanded that 'Christian themselves' were 'being seduced by these heathenish notions about an intelligent world'. As he noted in *The Sceptical Chymist*, 'divers chymical notions about matters philosophical are taken for granted and employed, and so adopted by very eminent writers'. The notion of 'spirit' in particular was applied 'to so many differing things, that this various and ambiguous use of the word seems to me no mean proof, that they [i.e., chemists and physiologists] have no clear and settled notion of the thing'.[61] Boyle further stated his objection to the philosophical imagery of spirits in his correspondence with Cambridge divine, Henry More, one of the most notable contemporary advocates of the 'spirit of nature'. Although Boyle acknowledged More's pious intentions, he questioned 'the Doctor' for 'proving the existence of an Incorporeal substance', since other persons 'of far less learning than He' could be misled by such a 'precarious Principle'.[62]

Thus, the philosophical discourse on the nature of things was portrayed by Boyle, and later on by Locke and Newton, as an institution that undermined the public good even when it was supported by scholars whose personal intentions were perhaps harmless. Rooted in academic life for centuries, it dominated the public discourse of scholars and corrupted the minds of ordinary people. Lay persons could barely understand the intricate definitions philosophers proposed but were nevertheless expected to acknowledge their authoritative position. The true path to God that had been originally shown by prophets and sages was thus gradually deserted by a profession that claimed to lead the search for truth.

---

pp. 42, 264-5). Thus, Newton apparently insisted upon the radical distinction between God and the world only later in his career, when he drafted the General Scholium for the *Principia*. Boyle's *Free Enquiry* was by then published, and it seems that Newton consulted this work, since he used Boyle's phrase, 'the soul of the world', in drawing the distinction between the creator and his creatures, as Boyle had stressed.

61 Boyle 1772, i, p. 609; Clericuzio 1994.

62 Boyle 1772, iii, 627-8. For detailed studies of the debate between Boyle and More on the relation between theology and experimental philosophy, see Shapin and Schaffer 1985, pp. 207-24; Henry 1990.

Experimental philosophy reclaimed the search for God's rule by offering an alternative to the idolatrous discourse on nature. As Boyle professed in his *The Christian Virtuoso*, there were on the one hand 'those, that cry up abstracted reason,' who, 'as if it were self-sufficient, exalt it in words', and, on the other, 'we that address reason to physical and theological experience' who 'exalt in effect'.[63] In contrast to obscure arguments, the experimental study of matters of fact provided the means by which every person could understand how God's rule applied to his or her own experiences. Newton similarly noted in a draft for the final query of his 1706 edition of *Opticks* that 'the argument wch all men are capable of understanding & by wch the belief of a deity has hitherto subsisted in the world is taken from the Phaenomena. We see the effects of a deity in the creation & what are his properties belong to experimental Philosophy'.[64]

Explanation and the Cultivation of Experience

Boyle's critique of idolatrous notions of nature set a new and arguably unprecedented constraint on philosophical solutions as applied to problems of causal explanation. Before Boyle, philosophers commonly associated natural order with natural kinds of phenomena. Changes in the physical world were accordingly attributed to causes and causal processes that were construed in terms of the properties of particular natures. Boyle's conception of true worship, by contrast, imposed the dichotomous distinction between the properties of physical bodies on the one hand, and God's rule on the other. Changes in the physical world were the mandate of God and not simply manifestations of essential properties of bodies.

Boyle's critique of idolatrous notions of nature primarily addressed the origins and early phases of the tradition of natural philosophy. He outlined the principal implications of his critique, though, in reference to the mechanical philosophy that dominated his early studies of the 1640s and 50s. Boyle never questioned the superiority of the new philosophy over the scholastic philosophy that was associated with the legacy of Aristotle. 'The schools', he noted, 'superficially' addressed the 'motion of bodies', as

---

63 Boyle 1772, v, p. 540.

64 Quoted from Schaffer 1983, p. 4. Newton, moreover, sought to apply his conception of true worship to the mathematical study of divine creation, as shown by Guicciardini 1999 and Ramati 2001.

against 'us', who set out 'to resolve the phenomena of nature into matter and local motion'.[65]

Historians have commonly noted that despite his commitment to the general tenets of mechanical philosophy, Boyle did not partake in the endeavor to develop a system of explicated principles of matter and motion. From the perspective on science as knowledge, his approach has been interpreted as an expression of his search for theories that derived their credibility from empirical evidence rather than metaphysical principles. Boas Hall, for example, suggested that Boyle developed a 'mechanical method for the explanation of forms, more mechanical than that of Gassendi, less complex and systematic than that of Descartes, better integrated with experimental philosophy than either'.[66]

As Chalmers convincingly demonstrated, Boas Hall overestimated the coherence of Boyle's mechanical method, which was too general and imprecise to be tested by his experimental studies. According to Chalmers, 'Boyle was a successful experimental scientist because he departed from the mechanical philosophy and Descartes wasn't because he didn't!'.[67] Offering a more historical interpretation, Sargent suggested that Boyle deliberately refrained from formulating a rigorous system of philosophical principles because he espoused a methodological strategy that encouraged eclectic and pluralist philosophical discourse from which the results of empirical trials formed the only admissible basis for selecting between explanations.[68]

---

65 Boyle 1772, ii, p. 2. In Boyle's writings, the principles of matter and motion are presented as the foundations of the 'mechanical' philosophy as well as the 'corpuscular' philosophy. While it appears that Boyle associated the latter primarily with chemical research, as pointed out in Clericuzio 1990 and Newman 1996, Anstey has convincingly shown that Boyle often used the two titles as synonyms (Anstey 2000). For example, Boyle stated in *The Excellency of Theology*, which he wrote in 1665 and published in 1674, that 'the corpuscular or mechanical philosophy strives to deduce all the phenomena of nature from adiaphorous matter, and local motion' (Boyle 1772, iv, p. 19). Since the examination in what follows does not directly pertain to the differences between Boyle's physical and chemical studies, it refers for convenience only to the mechanical philosophy.

66 Boas Hall 1958, pp. 92-3.

67 Chalmers 1993, p. 559.

68 Sargent 1995.

From the perspective on science as causal explanation, however, these interpretations seem to have put the cart before the horse. As shown throughout the previous chapters, philosophers traditionally taught that empirical evidence was a crucial component in the development of reliable explanations. They concomitantly considered the development of a coherent system of explanatory principles as a crucial component of their vocation. Philosophical coherence and empirical adequacy were never considered to be conflicting values. By their own methodological standards, philosophers did not fail to relate their causal explanations to empirical evidence. Conversely, Boyle could only render mechanical philosophy more empirical by revising its mode of explanation. Thus, because the common aim of philosophers was causal explanation, they construed solutions to the problem of knowledge as dependent on solutions to the problem of causal explanation. Historical studies that are based on the central dogma of science studies fail, therefore, to adequately interpret how natural philosophies differed from one another. Boyle's novel approach to empirical research was the outcome of his alternative solution to the problem of explanation, rather than the other way round.

Boyle embraced the principles of matter and motion as forming the most adequate framework for the analysis of natural phenomena. He nevertheless recognized that the philosophical endeavor to transform these principles into a causal account of the phenomena failed to conform to the rules of true worship that guided his approach to the study of the physical world. He therefore endorsed the mechanical philosophy in assuming that interactions between material corpuscles provided general analytic models of natural phenomena, but rejected the philosophical assumption that the properties of the interacting corpuscles could be considered to be the causes of the phenomena.

The attempt to accommodate the mechanical philosophy to the requirements of true worship led Boyle to draw a dichotomous distinction between the natural passivity of physical bodies on the one hand, and the divinely ordained conditions in which passive bodies interacted with one another on the other. Applying the principle of sufficient reason to his revised version of the principles of mechanical philosophy, Boyle thus noted that 'the action of God is sufficient to all the motions that occur among bodies'.[69] Although Boyle did not develop a specific concept of inertial motion, he stressed that adequate explanations ought to account for

69 Quoted from Anstey 2000, p. 211.

the state of physical bodies as possible outcomes of change rather than as expressions of their inherent dispositions to act in one way rather than another.

> The chief thing, that inquisitive naturalists should look after…is not so much what the agent is or does, as, what *changes* are made in the patient, to bring it to exhibit the phaenomena, that are proposed; and by what means, and after what manner, those changes are effected.[70]

The imagery of physical bodies as patients highlighted Boyle's intention to portray the source of changes in the physical universe as external to these bodies.

Boyle accordingly professed that researchers ought to be critical of the common conception of the impact of one body on another as a causal explanation of motion. It misleadingly suggested that 'one body does really and truly move another'. The empirical evidence demonstrated only the coincidence of motions: 'upon a moved body's hitting another, there follows a motion in the body'.[71] The aim of research was to explain this coincidence. The vulgar attribution of causal efficacy to physical bodies explained away the phenomenon. In his *Christian Virtuoso*, Boyle compared the researcher to 'an able judge' who was trained to keep vulgar notions of action at bay, and could thus pursue his calling as if he

> comes to hear and decide causes in a strange country. For the general notions he brings with him, and the dictates of justice and equity can give him but a very short and imperfect knowledge of many things...; and before he has heard the witnesses, he may be very apt to fall into prejudicate opinions of things.[72]

Boyle did not construe of the passivity of matter as applying exclusively to the macroscopic domain. As Clericuzio convincingly demonstrated, it was one of his principal considerations in developing his views on chemical elements.[73] In accordance with the critics of peripatetic explications of elementary principles, Boyle claimed that every form of matter was a compound of elementary particles whose properties

---

70 Quoted from Gillispie 1990, p.106; italics added.

71 Quoted from Anstey 2000, p. 211.

72 Boyle 1772, v, p. 539.

73 Clericuzio 1990.

determined the properties displayed by chemical analysis. These particles were therefore the minimal natural components, or 'minima naturalia', that characterized forms of matter of similar chemical properties. Although Boyle acknowledged the contribution of ancient atomists to the development of modern theories of chemical elements, he insisted that they and their modern disciples were erroneous in considering motion as an essential property of elementary particles. Alluding to Descartes' definition of matter, Boyle thus noted that its essence consisted of extension rather than motion, 'for Matter is no less matter when it rests, then when it is in motion'.[74]

The revival of doctrines of atomism in the 17th century appears to have been one of many targets of Boyle's critique of idolatry. Modern scholars often qualified their adherence to the ancient doctrines of Epicurus and Lucretius by claiming that atoms were the creation of God. However, they proposed their definitions as foundations of the causal explanation of natural phenomena, and thereby implied that the object of human understanding was the atom rather than God's government. Bacon, for example, considered the peripatetic doctrine of substantial forms as one of the 'Idols of the human mind,' in contradistinction to the idea of the 'atom' that was 'impressed' upon and 'defined' matter by 'the creator.' The atom was 'a true being, having matter, form, dimension, place, resistance, appetite, motion and emanations; which ... remains unshaken and eternal'.[75] Walter Charleton similarly professed, in his exposition of Gassendi's rehabilitation of the Epicurean philosophy in accordance with Christian theology, that 'God' invigorated or impreagnated' the atoms 'with an Internal Energy, or faculty Motive, which may be conceived the First Cause of all Natural Actions, or Motions...from the moment of its infusion, to the expiration of the Natures lease'.[76]

Boyle did not explicitly label modern definitions of the atom as forms of idolatry. These definitions nevertheless exemplified the tendency of Christian philosophers to construe causality as embodied in natural properties that were created by God but nevertheless gave rise to phenomena independently of God. Advocates of atomism acknowledged God's creation; yet they practically worshiped the atom instead of God. In his *Free Enquiry*, Boyle associated this concept of natural causality with

74 Boyle 1772, ii, p. 42.

75 Bacon 1961, v, p. 492.

76 Charleton 1654, p. 126.

the idolatrous notion of nature as 'God's vicegerent'. In his *Christian Virtuoso*, Boyle pointed out that deists typically renounced 'God's special providence' by claiming that 'after the first formation of the universe, all things are brought to pass by settled laws of nature'.[77] In *Free Enquiry*, Boyle further explained that the notion of laws of nature in natural philosophy was misleading because physical bodies were devoid of intelligence necessary for 'acting according to the declared will of a superior'.[78] This notion was applicable to the explanation of natural phenomena only when it denoted powers that determined law-like behavior of physical bodies.[79] The rules of true worship implied, moreover, that these powers ought to be construed as attributes of God's rule. As Boyle stated in *The Origin of Forms and Qualities*, 'God, by his general and ordinary support, preserves matter and motion from annihilation'.[80]

Thus, the relation between God and the material world undoubtedly played a critical role in Boyle's philosophy, and has become the subject matter of a considerable historical debate. While historians have commonly concluded that Boyle consistently stressed the dependence of the world on God's power, different and to some extent conflicting interpretations of his view of this dependence have been offered. For example, McGuire has suggested that Boyle considered God's intervention in the physical world as a necessary condition for its orderliness.[81] According to Jacobs, Boyle construed the relationship between the world and God as a quasi-legal idiom of consent to God's will as a creator and concurrence with His potential to exercise power.[82] Situating Boyle's position in the context of late medieval debates about divine providence, Shanahan interpreted it as claiming that both matter and God effectively contributed to causal processes in the physical world.[83] Henry suggested that Boyle attributed certain powers to divinely created entities to combat the atheistic implications that were associated with the principles of rational-mechanics in the philosophies of Hobbes and Descartes.[84] According to Anstey, Boyle

---

77 Boyle 1772, v, pp. 520-1.
78 Ibid., pp. 170, 175, 251-2, 521.
79 Ibid., p. 169.
80 Ibid., iii, p. 48-9.
81 McGuire 1972.
82 Jacobs 1994.
83 Shanahan 1988.
84 Henry 1986.

considered some forms of motion as inherent capacities of material bodies but attributed the overall order of the universe to the work of the divine lawmaker.[85]

The difficulty of reaching a conclusive interpretation of Boyle's position stems, in part, from Boyle's writings, which addressed the relationship between God and the material world on many different occasions. The problem of interpretation is related, moreover, to a more fundamental historiographic problem. The general assumption underlying the historical interpretations is that Boyle's theological position was somehow related to his scientific work. Yet in the absence of a clear view of the relationship between theology and science in Boyle's work, and especially of the scientific problem that Boyle sought to solve by means of his inquiry into the relationship between God and the material world, it is difficult to ascertain what his theological inquiry aimed to achieve.

It appears that the perspective on science as knowledge has undermined the historical attempt to interpret Boyle's views on divine government by obscuring the relationship between science and theology. Empirical knowledge is not directly affected by theological views. If Boyle's aim was scientific knowledge – as the central dogma of science studies implies – then it is indeed extremely difficult to examine how he brought his views on the relationship between God and natural causality to bear on his scientific work; hence the difficulty in ascertaining what his theological views precisely were. It is therefore not surprising that historical interpretations of Boyle's theological views have not directly addressed his scientific work. Those that have focused on his scientific work have not been directly concerned with his theological views.

The perspective on science as causal explanation offers a different approach to the problem of interpreting Boyle's theological inquiry. It has been demonstrated above that Boyle consistently related his theological views to the practical problem of developing a mode of explanation in natural philosophy that would agree with his deep concerns with the problem of worship. Since the problems of worship and of explanation were practical, Boyle did not find it necessary to engage in a rigorous explication of the relationship between God and the material world. His comments on this issue on many different occasions suggest that it was of critical importance to him. They suggest, however, that he did not seek to resolve the issue by elaborating on his comments to create a coherent

85 Anstey 2000, pp. 158-82.

philosophical doctrine. What he needed and indeed sought to do was to transform the notion of God's rule over the material world into a practical idiom that could form a critical component in his model of causal explanation.

As Boyle consistently proposed in his writings at least since the late 1650s, God's rule in the universe could be construed as embodied in the configurations of, or relations between, material bodies. Assuming that matter could not intentionally acquire properties that were manifested in the phenomena, Boyle concluded that these properties were systemically generated by the interactions between 'the parts of matter'. These interactions were causally effective as well as regular, thereby displaying the power and wisdom of divine government. As he proclaimed in his *The Origin of Forms and Qualities*, 'the wise Author of nature did not only *put matter into motion*, but, when he resolved to make the world, did so regulate and *guide the motions* of the small parts of the universal matter as to reduce the greater systems of them into the order they were to continue in'.[86]

Thus, Boyle's solution to the problem of subordinating natural philosophy to the rules of true worship primarily consisted of inverting the causal order that was commonly stipulated by mechanical philosophers. The latter commonly professed that the mechanical properties of particular natures determined the interactions between bodies. Boyle claimed, by contrast, that the interactions between bodies determined their distinctive properties.

The inadequacy of mechanical doctrines was metaphorically illustrated by explanations that suggested that clocks differed from one another because the parts with which they were made were essentially different. The craftsmen who designed and built clocks knew better. Clocks differed from one another because their elementary parts were configured in different ways. The study of the relationship between the parts was much more critical to the understanding of phenomena than the study of individual parts. The inadequacy of the teachings of philosophers stemmed from their common endeavor to consider properties of things in themselves, rather than in the context of their interrelations. For this reason, they failed to 'duly perpend, how great a difference there may be between a body

---

86 Boyle 1772, iii, p. 48.

considered absolutely, or by itself, and the same body considered in such circumstances, as it may be found in'.[87]

The dogmatic association of causality with the nature of things brought mechanical philosophers to depreciate God's rule. Descartes and his followers thus assumed that once God 'put matter into motion and established the laws of that motion', divine government was no longer required 'for the production of things corporeal'. The advocates of atomism similarly assumed that once God created atoms, they could on their own 'bring mere matter into so orderly and well contrived a fabric as this world'.[88]

By inverting the causal relation between parts of matter and their configurations in space and time, Boyle did more than simply propose a mode of explanation that better accorded with true worship. The new mode of explanation provided empirical and experimental research the critical weight of which it hitherto lacked in the philosophical study of natural phenomena. Empirical research was obviously the most effective means of identifying the configurations of material objects, to ascertain how they changed, and to assess the effects of the changing configurations. Experimental trials were considerably more critical to Boyle's endeavor to figure out configurations than they were to traditional endeavors to define the nature of things. Thus, Boyle's experimental philosophy was the outcome of his endeavor to solve the problem of worship by revising mechanical philosophy.

Boyle illustrated the relationship between worship, explanation, and experimentation by means of a lock metaphor, which he presented in his *Origin of Forms and Qualities.*[89] The metaphor conveyed Boyle's belief that the physical universe functioned as system of devices whose practical value for God's creatures formed integral aspects of His benevolent rule. The study of these devices was therefore a religious duty that also offered an opportunity to acquire the tools with which human agents could better accommodate themselves to God's dominion.

From the viewpoint of Boyle's revision of mechanical philosophy, the physical process of locking (or unlocking) involved a dynamic system of two principal material units, namely a lock and a key. This system could model interactions between material entities that generated mechanical as

---

87 Ibid., v, pp. 176, 248-9.

88 Ibid.

89 Ibid., iii, p. 18.

well as chemical effects. When the key and the lock were examined each on its own, they appeared to be endowed with several properties which Boyle earlier in this treatise classified as 'catholic', namely material 'bulk, figure' (or, 'bignes' and 'shape', respectively), 'and either motion or rest'. Following Descartes, Boyle suggested that these properties were necessarily associated with the idea of matter.[90] Although the specific magnitude of each of these properties was variable, the bodies were passive, or inert, and could not on their own change these magnitudes.

Thus, neither the properties of the key nor those of the lock could be construed as sufficient for explaining the effects of the process that was induced through their interaction with one another. As Boyle further stressed, the process did not add 'any *real or physical entity* either to the lock or the key, each of them remaining indeed nothing but the same piece of iron, just so shaped, as it was before'. To explain the process, it was therefore necessary to attribute a 'capacity', 'faculty', or 'power', to the system of the interacting parts rather than to any of the individual parts themselves.[91] As noted earlier, Boyle claimed in his *Usefulness* that power and structure were the two basic dimensions of God's dominion. The metaphor illustrated how he transformed the theocentric image of the universe into an image based on the mechanical principles of matter and motion.

The metaphor further demonstrated the critical role practical experience played in Boyle's solution to the problem of causal explanation. However philosophically learned they were, the inexperienced spectators who observed the key and the lock could hardly realize that these objects were parts of an operative system, and could not possibly understand how it operated. Cultivating practical experience was the key to successful learning that aimed to figure out how the system worked and how it could be applied to the production of goods. Boyle's philosophy was experimental because he considered methodical empirical research to be the crucial means to making sense of empirical evidence.

Boyle accordingly distinguished the experimental philosopher from 'a libertine, though ingenious, or a sensualist, though curious', or 'a mere empiric, or some vulgar chemist...who too often makes experiments, without making reflection on them, as having it more in his aim to produce effects, than to discover truths'. 'The person I here mean, is such a one, as

90 Ibid., v, 35.

91 Ibid., iii, pp. 18, 21; italics added.

by attentively looking about him, gathers experience...on which he is disposed to make such reflection, as may (unforcedly) be applied to confirm and encrease in him the sentiments of natural religion, and facilitates his submission and adherence to the Christian religion'.[92] By observing physical objects, the researcher aimed to identify empirical properties whose interrelations revealed the ordained devices that made up God's dominion.

Observation provided the groundwork for the design of experiments that could reproduce valuable correlations between empirical properties. Boyle thus advised 'that the mind do, as it were, walk round the object it is to contemplate, and view it on every side, observing what differing prospects it will that way afford'; just as 'a painter, or an anatomist', before embarking upon his creative work, 'looks upon a man's body, first when the face and belly are towards him, then when the back and other hinder parts are so'.[93] Rather than merely a means to obtain evidence for or against a particular theoretical claim, observations were intended to be part of the creative endeavor to transform the evidence into a coherent representation of the 'keys and locks' of the physical universe.

Empirical research was therefore valuable in as much as it was integrated with the relatively more abstract aspects of causal explanation. Conversely, ideas and conjectures about causal relations between physical objects were valuable if they could be brought to bear on the elaboration of specific strategies of empirical research. Philosophers routinely demonstrated that their hypotheses about the nature of things were consistent with empirical evidence; yet these hypotheses were often too abstract or speculative to guide empirical research. An 'excellent hypothesis', by contrast, enabled 'a skillful Naturalist to foretell future Phenomena, by their Congruity or Incongruity to it; as Things that ought or ought not to be Consequent to it'.[94]

Boyle's new mode of explanation could thus be construed as involving a trade-off between philosophical clarity and empirical robustness of explanations, in which the emphasis was transferred from the former to the latter. Boyle did not deny that 'it is an advantage as well as a satisfaction, to know in general, how the qualities of things are deducible from the primitive affections of the smallest parts of matter'. However,

92 Ibid., v, p. 524.

93 Ibid., iv, p. 597.

94 Quoted from Westfall 1956, p. 117.

since philosophers had until then failed to demonstrate such deductions, he confidently concluded that

> whether we know that or not, if we know the qualities of this or that body they compose, and how it is disposed to work upon other bodies, or be brought on by them, we may, without ascending to the top in the series of causes, perform things of great moment, and such as, without the diligent examination of particular bodies, would, I fear, never have found out a priori, even by the most profound contemplator.[95]

As Boyle noted more succinctly in his *Commonplace Book*, 'the most useful notions we have, both in Physics, mechanicks, Chymistry, & ye medicinal art, are not deriv'd from ye first principles, but from intermediate Theorys, notions and rules'.[96]

Boyle's preference for middle-road explanations that derived their meaning primarily from their function as tools for the factual relation of empirical properties, and from the experimental reproduction of effects, were the hallmarks of his experimental philosophy. As shown in the previous chapters, Locke and Newton consistently sought to demonstrate that experimental philosophy offered a model of philosophical learning that was undeniably superior to the search for a definition of the nature of things.

Historical studies have shown that Spinoza and Hobbes were among the first philosophers who criticized Boyle for proposing explanations that failed to conform to the philosophical ideal of defining the nature of things.[97] Neither of these critics underestimated the value of observation or experiment. Hobbes claimed, for example, that Gilbert's experimental reports in *De Magnete* set a new standard of rationality in the philosophical study of magnetism and electricity.[98] He considered Boyle's pneumatic experiments, by contrast, to be both insufficiently coherent and decisive for resolving the philosophical question of the vacuum. Spinoza attacked Boyle's chemical studies of niter, and similarly claimed that the experimental reports of the latter failed to provide sufficient empirical evidence that demonstrated his chemical analysis.

---

95 Boyle 1772, i, pp. 198-9.

96 Quoted from Clericuzio 1990.

97 Hall and Hall 1964; Shapin and Schaffer 1985, esp. pp. 80-154; Clericuzio 1990.

98 Hobbes 1840, vii, p. 57.

As Hobbes pointed out in his *Physical Dialogue*, explanations that derived their meaning from the experimental production of effects lacked the methodical coherence that philosophers traditionally sought by predicating their explanations on the explication of the nature of things. Alluding to the Royal Society, he noted with disapproval that 'this new Academy abounds with most excellent ingenious men. But ingenuity is one thing and method is another'.[99]

As shown in Chapter Four, Newton encountered similar opposition when he published his new theory of the heterogeneous composition of light. His critics acknowledged the novelty of his experimental work, but found his demonstrations lacked the methodical structure that characterized discourse on the nature of light and colors. Critics of experimental philosophy failed to recognize the extent to which it transformed the means of achieving their common aim, namely the causal explanation of natural phenomena. They appraised its accomplishments in light of the traditional search for the nature of things, rather than in accordance with its innovative endeavor to understand the relationship between the empirical properties of physical bodies.

By labeling experimental philosophers as ingenious, Hobbes nevertheless singled out highly significant characteristics of their calling and its cultural origins. In mid 17th century England, 'Ingenuity' was commonly associated with the creativity and cleverness of preachers as well as craftsmen.[100] Both of these professions aspired to excel in transforming resources – in their respective domain of expertise – into publicly desirable goods. Philosophers traditionally claimed that methodical cultivation of rationality distinguished their vocation from other professions. The experimental philosophers' new method of explanation was highly innovative in part because it was based on the application of techniques that were applied by preachers and craftsmen to the domain of natural philosophy.

99 Hobbes 1985, p. 347.

100 Houghton 1942; Greene 1981; Shapin and Schaffer 1985, pp. 128-31.

Chapter Eight

Science as an Institution of Human Understanding: Conclusions

Historical Aspects

This book has unfolded the story of one of the most radical and consequential changes in the history of early modern science. It was radical because it marked out the break with the discourse on the nature of things that dominated the philosophical study of the physical world since its inception in ancient Greece. In this respect, this change differed, for example, from the impact of mathematical tools or mechanical images and models on science, which evolved within the philosophical discourse on the nature of things since the rise of humanism in 14th century Italy. It had far-reaching consequences because it transformed empirical research into a crucial and most vital means of explaining natural phenomena. The break with age old philosophical tradition and the changing roles of empirical research were inseparable. Both were the result of establishing a new mode of explanation proposed as an alternative to the traditional search for a definition of the nature of things. The alternative, in a nutshell, consisted of understanding correlations between empirical properties.

Explanations in natural philosophy always relied on empirical evidence. However, as long as these explanations were elaborated on in relation to the definition of the nature of things, they were practically separated from the tools of empirical research. Gathering empirical information was one thing. Demonstrating how definitions could be brought to explain it was another. This division of labor structured the institution of natural philosophy as a profession. It was routinely the common endeavor of philosophers to develop different explanations of similar empirical evidence or phenomena on the basis of different definitions of the nature of things.

The new endeavor to demonstrate how empirical properties of physical objects were correlated effaced the division between empirical and philosophical studies. The result was a major cultural change that pertained, in particular, to the cultivation of two fundamental human capacities, namely thought and experience. Two novel modes of combining thought with experience became particularly notable. The first pertained to the content of factual judgments. The received philosophical view suggested that a statement of fact conveyed the experience of a contingent state of affairs, and therefore differed from the explication of its causes. The alternative approach to the problem of explanation demonstrated, by contrast, that factual judgments included a means of transforming empirical evidence into intelligible representations of natural processes. The construction of such representations aimed to reveal new facts with meanings that denoted causal relations. For this reason, the discovery of new facts could convey solutions to philosophical and practical problems.

The second cultural innovation pertained to the practice of experimentation. Philosophers traditionally assumed that experiment was valuable as a means to obtain empirical evidence relevant to proper definitions of the nature of a particular object. For this reason, their work primarily pertained to evidence rather than to the experimental means of obtaining it. With a new mode of explanation, experiment became an explanatory practice; a technique for making sense of evidence. The effectiveness of this technique was a function of the quality of the design and application of experimental instruments, apparatuses, and procedures. Hence, the construction of representations of physical processes that explained phenomena involved as much engineering as it did theoretical thinking.

These two cultural innovations were closely related to one another. Together they provided a framework for integrating causal explanations, techniques of experimentation, and factual representations of empirical evidence within unified research products that could be disseminated in the scientific community. As shown in the previous chapters, however, this novel type of research product transformed the tradition of natural philosophy through the effort of neither a single person nor of a group of persons working in collaboration with one another. Rather, the historical transformation happened gradually, as a result of the independent initiatives and distinctive contributions of several researchers who could not foresee the consequences of their own creative endeavors.

The most significant effect of the new mode of explanation on the history of science was initially manifested in the emergence of the new discipline of electrical research that followed Gray's discovery of electrical conductivity. Gray did not include in his published work programmatic statements that addressed the development of the science of electricity. The most critical factor in the emergence of the new discipline was electrical conductivity itself. As an explanatory construct, a technique of experimentation, and an apparently undeniable fact, electrical conductivity set the basic constraints and opportunities that directed the projects and careers of new generations of researchers. Within a few decades, the new discipline expanded far beyond what had been previously achieved by natural philosophers for centuries.

The mode of explanation that transformed electrical research in the 1730s and 40s was originally proposed by Boyle as an alternative to the philosophical discourse on the nature of things in the 1650s and 60s. Unlike Gray, Boyle intended for a great deal of his publications to be manifestos of the new way of philosophical research. Boyle hoped to see the emergence of research communities devoted to advancing his own experimental studies in chemistry, light and colors, the spring of air, or the other subjects he studied. His hope did not, however, materialize. Few scholars followed his lead to contribute to the consolidation of his experimental philosophy. By historical coincidence, among his disciples were Locke and Newton, whose contributions were highly individualistic and immediately aroused great interest. Their intellectual endeavors were much more focused than Boyle's, and enabled them to develop more systematic and specific demonstrations of the new mode of explanation.

It appears, in retrospect, that Boyle's greatest contribution consisted of elucidating different aspects of the calling of experimental philosophy in a wide variety of cultural and historical contexts. His numerous writings on the history of religion, the history of natural philosophy, theology, morality, intellectual faculties, mechanical philosophy, and a wide variety of topics in natural philosophy and practical sciences, created a culturally rich and complex portrayal of his new initiative. This literary culture was the foundation of Locke's and Newton's broad conceptions of experimental philosophy and its mission to change the history of philosophy.

The Royal Society of London provided an important institutional setting in which endeavors to consolidate the new mode of explanation gained public recognition as a collective undertaking. The Society was England's most prestigious organization for the advancement of knowledge, and was widely associated with the ideals of cooperative and experimental learning. The experimental philosophy gained credence as a public enterprise in part because its advocates channeled their efforts through the Royal Society.

However, as a creative endeavor to transform philosophical understanding of the physical world, experimental philosophy was not the invention of the Society's fellows. Boyle's conception of its calling originated in studies he conducted during the late 1640s and 1650s, before the Society was founded. Newton and Locke committed their careers to advancing the new mode of explanation before they were elected fellows of the Society. The Society was eager to give official recognition to achievements that agreed with its mandate, such as Newton's and Gray's discoveries, which enhanced its own prestige as a voluntary association devoted to servicing the public good.

The study presented in this book has aimed to demonstrate that the experimental philosophy of Boyle, Locke, and Newton transformed the mode of explanation that traditionally characterized the profession of natural philosophers. It has not aimed to systematically explain why these individuals decided to develop this particular alternative. It nevertheless appears that their innovative works were predominantly influenced by the educational culture of Christian humanism. This culture preached religious piety that was highly secular in its focus on the moral integrity of believers in private and public spheres of action. It stressed the importance of methodically rigorous but theologically moderate examinations of the scriptures, and emphasized their practical uses. By sanctifying disciplines of conduct in everyday life, it empowered the values of the production and public transaction of tangible goods. Its institutional affinities with movements of religious reform encouraged experiments in social and political change.

Christian humanism was imported to England by Protestant clergymen in the late 16th and early 17th centuries. Though aiming primarily to reform the Church of England, these clergymen disseminated ideas and ideals that greatly appealed to a lay public whose interests were predominantly secular, and who was powerful enough to redirect the movement of religious reform toward secular institutions. During the first half of the 17th century, however, the use of cultural resources that had earlier originated in the humanistic movement on the continent was overshadowed in England by political struggles between religious factions. The Restoration period brought about a new climate of greater civil peace and stability, in which these resources were innovatively applied to establishing experimental philosophy.

Historiographic Aspects

The historical examination of the works of Boyle, Locke, and Newton has been conducted predominantly within the historiographic framework of the Scientific Revolution. Pioneered during the first half of the 20th century, this framework has been designed to illuminate changes in the practice of the study of nature in the early modern period, and especially to explain the rise of modern science. In accordance with the central dogma of science studies, however, historians have commonly assumed that the principal aim of empirical research in the natural sciences has always been the acquisition of knowledge. Their studies have accordingly been predicated on the assumption that the aim of empirical research did not change during the Scientific Revolution.

These historiographic assumptions stand squarely against the experimental philosophy's center of gravity, namely the relationship between empirical research and philosophical explanation (rather than knowledge in general). Historical studies have consequently failed to adequately examine, for example, how Boyle's experimental philosophy followed his critique of idolatrous notions of nature; his impact on Locke and Newton; the difference between Locke's and Hobbes' or Gassendi's views on the relation between experience and the intellectual faculties; the difference between Descartes' and Newton's empirical research on spectral colors; Newton's role in Gray's discovery of electrical conductivity; or the difference between Gilbert's and Gray's study of electrical phenomena. In general, historians have failed to recognize that the advocates of experimental philosophy radically transformed the philosophical aim of empirical research.

From a historiographic perspective on science as causal explanation, the study of the experimental philosophy of Boyle, Locke, and Newton has developed along the following assumptions: modern scientists invest much greater amounts of resources in the development of empirical research than their predecessors did before the 17th century. Since the aim of natural scientists has always been causal explanation, it follows that modern scientists consider empirical research as much more vital to the search for their causal explanations than their predecessors in ancient times or in the late Middle Ages. Assuming that the latter knew what kind of empirical research was most suitable to the search for causal explanations, it follows that the radical change in the value of empirical research was the result of a proportionally radical change in the mode of causal explanation. Compared with predecessors before the 17th century, modern natural scientists seek causal explanations that more heavily rely on empirical research. Since the philosophical works of Boyle, Locke, and Newton have always been known for their emphasis on the value of empirical research, the historical question has been how they contributed to the rise of a new mode of explanation that more heavily relied on empirical research.

The central dogma has failed, moreover, to provide an adequate historiographic perspective on Newton's discovery of the heterogeneous composition of light and Gray's discovery of electrical conductivity. The central dogma claims that the aim of science is knowledge. Historical evidence suggests, however, that scientists define research problems in different ways and accordingly develop different research projects to solve them. These variations do not support the assumption that scientists commonly follow a shared conception of knowledge.

Following the central dogma, historians have assumed that the aforementioned discoveries were the result of the search for knowledge. However, since the central dogma claims that every scientist seeks to advance knowledge, its appears that scientific discoveries must be attributed to highly idiosyncratic characteristics of individual researchers. Historians have accordingly concluded that Newton discovered the heterogeneous composition of light because he was simply more talented than his predecessors, and that Gray discovered electrical conductivity because he was more fortunate than other researchers. Thus, historical studies have failed to adequately explain how Newton or Gray systematically developed research projects that led to novel theoretical conclusions.

An adequate historiographic perspective on scientific discoveries and other forms of change in the practice of science must be predicated on the assumption that the aim of science is a variable that scientists may define in different ways. Moreover, this variable must be in part subjective to allow equally reasonable scientists to define it in different ways for reasons that are acceptable to some of them but may not be acceptable to others. A partially subjective definition of the aim of science is not necessarily private. If it incorporates cognitive and cultural considerations in addition to its subjective aspect, it may comprise constraints on, and opportunities for, common initiatives to advance science.

The variable aim of scientific research that has been proposed in this book is causal explanation. This proposal is by no means novel. Scientists have always aimed to advance human understanding by elaborating causal explanations that are credible and persuasive. For this reason, they have always sought explanations that are logically coherent and empirically reliable. Scientists nevertheless develop logically coherent and empirically reliable explanations that may radically differ from one another. These differences explain, in turn, how researchers advance science in different directions. Scientific discovery appears, from this perspective, to be the result of the elaboration of novel explanatory practices. This creative process delineates the elementary historiographic framework for the reconstruction of scientific innovations.

In accordance with these assumptions, it has been suggested in this book that Newton's theory of light and colors and Gray's theory of electrical conductivity were neither more logically coherent nor more empirically adequate than the theories that were proposed by their respective predecessors. Their theories were innovative because they demonstrated new ways of transforming empirical evidence into causal explanations that solved important philosophical problems. The novelty of their theories consisted, more specifically, of rejecting the traditional quest for a definition of the nature of things, and proposing, instead, a causal explanation of correlations between empirical properties.

Sociological Aspects

Science is a social institution. Its practice requires and contributes to the cultivation of natural human capacities, namely thought and experience. The cultivation of these and other capacities generates culture. Cultures is comprised of constraints and opportunities that regulate human conduct; the satisfaction of needs and desires. Specific cultural constraints and opportunities define specific institutions. Understanding science as a social institution hinges, therefore, on understanding the specific cultural constraints and opportunities that regulate the pursuit of the aims of science.

Social studies of science have been predicated on the central dogma of science as the pursuit of knowledge. They have evolved primarily in relation to two mutually exclusive definitions of knowledge. Originally, sociologists adopted positivist definitions of knowledge as a product of methodological assessments of theories about phenomena. Positivist explications of scientific method have implied that the growth of scientific knowledge did not directly hinge on social interactions within the scientific community or between scientists and other groups in society. Positivist sociologists have accordingly focused their studies on the cultural contexts and social organizations in which the methods of science were allegedly applied.[1]

With the gradual demise of positivist definitions of scientific knowledge in the 1960s and 70s, social studies have rejected the notion that the production of knowledge is regulated by methodological principles. They have concomitantly abrogated the distinction between the philosophical study of scientific knowledge on the one hand, and the sociological study of the cultural or organizational contexts of scientific research. Claiming that science in its entirety is a social activity, these studies have proposed social definitions of knowledge as a product of negotiations and power struggles among scientists and other interest groups over the procedures and results of empirical research.[2]

This study has shown, however, that both positivist and social perspectives on scientific knowledge are fundamentally inadequate. Positivist definitions fail to take into account that the methodological assessment of theory presupposes the representation of empirical evidence; a product of creative endeavors that depend on available cultural resources. Scientific innovations in the representation of evidence presuppose the appraisal of prevailing representation in light of cultural values. Boyle's work, for example, demonstrates that the role of empirical research in ascertaining correlations between empirical properties was shaped by the values of Christian humanism. The historical evidence shows, moreover, that Boyle consistently endeavored to bring these cultural values to bear on practically every aspect of his research work. Ironically, positivist philosophers have embraced the ideal of scientific progress, though progress in science appears to have been achieved because scientists found innovative ways of using cultural resources as a means of transforming

---

1 See, for example, Merton 1973; Ben-David 1981, 1984.

2 See, for example, Latour and Woolgar 1979; Knorr-Cetina and Mulkay 1983; Collins 1985; Lynch 1985; Latour 1987; Woolgar 1988.

empirical evidence into novel representations of objects in the physical world.[3]

Although social studies have claimed to offer an alternative to positivist accounts of scientific practice, they have followed the teachings of positivism in assuming that cognition and social interaction are mutually exclusive dimensions of human conduct. Their portrayals of science as a social activity is in part an artifact of ignoring the intellectual and cognitive efforts invested in the search for valuable explanations. Contrary to social definitions of knowledge, the study of experimental philosophy suggests that scientists endeavor to bring social and cultural contexts to bear on their work, because they are interested in utilizing resources these contexts offer to develop their intellectual and cognitive skills.

From the perspective on science as an activity that aims to augment human understanding of the physical world, the dichotomy between cognition and social interaction appears fictitious. Human understanding hinges on cultivating thought and experience, which is a highly variable process. For this reason, the explanation of natural phenomena acquires a plethora of forms that vary especially across cultures and social institutions. Yet human agents are interested in taking part in public transactions of explanations. For this reason, they are engaged in the development of institutions that regulate the cultivation of thought and experience. The sociological study of human understanding and its historical evolution is therefore a study of both society *and* cognition, and intellectual work *and* culture.

The historical study presented in this book has addressed the rise of experimental philosophy from this broad sociological perspective. Anthropological and historical studies have shown that in many pre-industrial societies, understanding the natural environment has not been institutionally separated from activities pertaining to basic economic and political interests, such as gathering and producing food, maintaining health, or protecting the community from its enemies. The innovations of ancient Greek philosophers gave rise to relatively more specialized and autonomous institutions of learning within which natural philosophy as a profession evolved. Studying the nature of things was the principal interest

---

3 For studies on the roles of cultural values and models of action in innovating representations of empirical evidence in science, see, for example, Ben-David 1960; Bloor 1976; Shapin 1979, 1982; Freudenthal 1987; Jardine 1991; Wise and Smith 1986.

that shaped the institutional structure of this profession. This structure was in turn transformed with the rise of modern cultures of empirical research in the natural sciences. As shown in this book, this institutional change was engendered by the endeavor to cultivate thought and experience in ways suited to a new mode of understanding the physical world.

From a sociological viewpoint, then, scientific explanations are institutionally regulated cultural activities. They are based on empirical evidence and logical principles; yet they relate experience and logical thought to one another in specific modes that are shaped by cultural values and interests. For this reason, scientific explanations can neither be refuted nor proven by methods of knowledge. Their value varies as a function of their use in the specific contexts in which they are transacted. Like other cultural products, their value may increase in one context and decrease in another. They may provide the foundations of research in one discipline and be considered inadequate in another. Like other tools, they end their life-cycle when they are replaced by more useful inventions.

In examining scientific explanations as social institutions that regulate thoughts and experiences of individual scientists and their public transactions, the sociology of human understanding rejects the philosophical doctrines of scientific realism. The latter propose that scientific laws and theories describe an objective reality, a reality that is independent of human history and culture.[4] Scientific realism implies, for example, that Gray's achievement consisted of revealing a physical property, namely electrical conductivity, that exists independently of the work that was invested by Gray and his predecessors in sorting out and explaining electrical phenomena. The realist implication is self-defeating, however, because all that the philosophical realist could say about this property derives from the creative endeavor of scientists.

However, the sociological study of human understanding by no means underestimates the value of truth in scientific research. Following Locke's *Essay*, it espouses a notion of truth as a value that, in the context of the study of the physical world, may characterize beliefs rather than claims to knowledge. Beliefs belong to a person as a whole rather than to a

---

4 For a discussion of various approaches to philosophical realism in the context of empirical research in modern natural sciences, see Hacking 1983; Churchland and Hooker 1985. For a critique of philosophical realism in respect to the cognitive and explanatory values of scientific theories, see van Fraassen 1980, 1985.

disembodied mind. Their truthfulness hinges on their practical value. The truth value of beliefs does not therefore reflect a reality separated from human life. Rather, this value is an expression of the ineradicable dependence of human agents on tools of thought and action, the usefulness of which inevitably depends, in turn, on a reality that is greater than their human inventors.

Bibliography

Aaron, R.I., 1955. *John Locke*. Oxford: Clarendon Press.

Aarsleff, H., 1969. 'The state of nature and the nature of man in Locke', in J.W. Yolton (ed.), *John Locke – Problems and Perspectives*. Cambridge: Cambridge University Press, pp. 99-136.

Aiton, E.J., 1970. 'Ioannes Marcus Marci (1595-1667)', *Annals of Science*, 26: 153-64.

Alexander, P., 1985. *Ideas, Qualities and Corpuscles. Locke and Boyle on the external world.* Cambridge: Cambridge University Press.

Allen, D.E., 1976. *The Naturalist in Britain: A Social History.* London: A. Lane.

Altick, R.D., 1978. *The Shows of London.* Cambridge: Belknap Press.

Anstey, P.R., 2000. *The Philosophy of Robert Boyle.* London: Routledge.

Aquinas, St. T., 1974. *Summa Theologiae*, ed. and trans. T.C. O'brien, 61 vols. New York: Blackfriars.

Aristotle, 1992. *Aristotle De Anima,* trans. with an introduction D.H. Hamlyn. Oxford: Clarendon Press.

Arnauld A. and P. Nicole, 1967. *Port Royal Logic*, ed. H. Brekle. Stuttgart-Bad Cannstatt: Frommann.

Ashcraft, R., 1969. 'Faith and Knowledge in Locke's Philosophy', in J.W. Yolton (ed.), *John Locke: Problems and Perspectives*. Cambridge: Cambridge University Press, pp. 194-223.

Ashworth, W.B., 1990. 'Natural History and the Emblematic World View', in D.C. Lindberg and R.S. Westman, *Reappraisals of the Scientific Revolution.* Cambridge: Cambridge University Press, pp. 303-32.

—, 1986. 'Catholicism and Early Modern Science', in D.C. Lindberg and R.L. Numbers (eds.), *God and Nature. Historical Essays on the Encounter between Christianity and Science.* Berkeley: University of California Press, pp. 136-66.

Atherton, M., 1983. 'Locke and the Issue Over Inateness', in L.S. Cauman (ed.) *How Many Questions? Essays in Honor of Sidney Morgenbesser.* Indianapolis: Hackett, pp. 223-42.

Axtell, J.L., 1969. 'Locke, Newton and the two cultures', in J.W. Yolton (ed.), *John Locke: Problems and Perspectives.* Cambridge: Cambridge University Press, pp. 165-82.

Ayers, M., 1991. *Locke,* 2 vols. London: Routledge.

—, 1994. 'The Foundations of Knowledge and the Logic of Substance: The Structure of Locke's General Philosophy', in G.A.J. Rogers (ed.), *Locke's Philosophy. Content and Context.* Oxford: Clarendon Press, pp. 49-73.

—, 1998. 'Theories of Knowledge and Belief', in D. Garber and M. Ayers (eds.), *The Cambridge History of Seventeenth-Century Philosophy.* Cambridge: Cambridge University Press, pp. 1003-61.

Bacon, F., 1952. *Novum Organum,* ed. R.M. Hutchins. Encyclopaedia Britannica, Great Books of Western World.

—, 1961. *The Works of Francis Bacon,* ed. J. Spedding, R.L. Ellis, and D.D. Heath, 14 vols. Stuttgart: F. Frommann.

—, 1980. *The Advancement of Learning and New Atlantis*, ed. A. Johnston. Oxford: Clarendon Press.

—, 2000. *The New Organon*, ed. L. Jardine and M. Silverthorne. Cambridge: Cambridge University Press.

Baldwin, M.R., 1985. 'Magnetism and the Anti-Copernican Polemic', *Journal for the History of Astronomy*, 16: 155-74.

Barnes, B., 1974. *Scientific Knowledge and Sociological Theory.* London: Routledge & Kegan Paul.

Barnes, B. and S. Shapin (eds.), 1979. *Natural Order: Historical Studies of Scientific Culture.* Beverly Hills: Sage.

Barnes, B. and D. Bloor, 1982. 'Relativisim, Rationalism, and the Sociology of Knowledge', in M. Hollis and S. Lukes (eds.), *Rationality and Relativism.* Oxford: Basil Blackwell, pp. 21-47.

Beare, J.I., 1906. *Greek Theories of Elementary Cognition from Alcmaeon to Aristotle.* Oxford: Clarendon Press.

Bechler, Z., 1973. 'Newton's Search for a Mechanical Model of Colour Dispersion: A suggested Interpretation', *Archive for History of Exact Sciences*, 11: 1-37.

—, 1974. 'Newton's 1672 Optical Controversies. A study in the Grammar of Scientific Dissent', in Y. Elkana (ed.), *The Interaction between Science and Philosophy.* Atlantic Highlands: Humanities Press.

Ben-Chaim, M., 1998. 'Doctrine and Use: Newton's 'Gift of Preaching'', *History of Science*, 36: 269-298.

—, 1998a. 'Knowledge and the Body-Mind Problem & The Myth of the Framework by Karl Popper', Critical Study, *Philosophia,* 26: 529-544.

—, 1999. 'The Intention of Experiment', Essay Review, *Studies in History and Philosophy of Science*, 30: 173-181.

—, 2000. 'The value of facts in Boyle's experimental philosophy.' *History of Science* 38: 57-77.

—, 2000a. 'Locke's Ideology of 'Common Sense'', *Studies in History and Philosophy of Science*, 31: 473-501.

—, 2001. 'The discovery of natural goods: Newton's vocation as an 'experimental philosopher'', *British Journal for the History of Science,* 34: 395-416.

Ben-David, J., 1960. 'Roles and Innovations in Medicine', *American Journal of Sociology,* 55: 557-568.

—, 1981. 'Sociology of Scientific Knowledge', in J.F. Short (ed.), *The State of Sociology: Problems and Prospects.* Beverly Hills: Sage.

—, 1984. *The Scientist's Role in Society. A comparative Study.* Chicago: The University of Chicago Press.

Benjamin, A.E., G.N. Cantor, and J.R.R. Christie (eds.), 1987. *The Figural and the Literal: Problems of Language in the History of Science and Philosophy, 1630-1800.* Manchester: Manchester University Press.

Bennett, J., 1965. 'Substance, Reality, and Primary Qualities', *American Philosophical Quarterly*, 2:1-17.

—, 1987. 'Substratum', *History of Philosophy Quarterly*, 4: 197-215.

Bennett, J.A., 1986. 'The Mechanics' Philosophy and the Mechanical Philosophy', *History of Science,* 24: 1-28.

—, 1991. 'The Challenge of Practical Mathematics', in S. Pumfrey, P. Rossi, and M. Slawinski (eds.), *Science, Culture and Popular Belief in Renaissance Europe.* Manchester: Manchester University Press.

Biagioli, M., 1989. 'The Social Status of Italian Mathematicians, 1450-1600', *History of Science*, 27: 41-95.

—, 1993. *Galileo, Courtier: The Practice of Science in the Culture of Absolutism.* Chicago: The University of Chicago Press.

Birch, T., 1757. *The History of the Royal Society of London*, 4 vols. London.

Blair, A.M., 1985. *The Geometrical Optics of Isaac Barrow: The Scientific Epistemology of a Restoration Divine*, M.Phil. dissertation, Clare College, Cambridge.

Bloor, D., 1976. *Knowledge and Social Imagery.* London: Routledge and Kegan Paul.

—, 1999. 'Anti-Latour', *Studies in History and Philosophy of Science*, 30: 81-112.

—, 1999. 'Reply to Bruno Latour', *Studies in History and Philosophy of Science*, 30: 131-6.

Boas Hall, M., 1952. 'The Establishment of the Mechanical Philosophy', *Osiris*, 10: 412-541.

—, 1958. *Robert Boyle and Seventeenth-Century Chemistry.* Cambridge: Cambridge University Press.

—, 1976. 'Science in the Early Royal Society', in M. Crosland (ed.), *The Emergence of science in Western Europe.* London: Macmillan, pp. 57-77.

Boas Hall, M. and A.R. Hall, 1959. 'Newton's Electric Spirit: Four Oddities', *Isis*, 40: 473-6.

Boswell, J., 1964-71. *Boswell's Life of Johnson,* ed. G.B. Hill, revised and enlarged by L.F. Powell, 6 vols. Oxford: Clarendon Press.

Boyer, C.B., 1987. *The Rainbow From Myth to Mathematics*. Princeton: Princeton University Press.

Boyle, R., 1664. *Experiments and Considerations touching Colours.* London.

—, 1673. *Of the Great Efficacy of Effluviums.* London.

—, 1772. *The Works of the Honourable Robert Boyle*, ed. T. Birch, 6 vols. London.

—, 1991. *The Early Essays and Ethics of Robert Boyle,* ed. J.T. Harwood. Carbondale: Southern Illinois University Press.

—, 1996. *A Free Enquiry into the Vulgarly Received Notions of Nature*, ed. E.B. Davis and M. Hunter. Cambridge: Cambridge University Press.

—, 1999. *The Works of Robert Boyle*, ed. M. Hunter and E.B. Davis, 14 vols. London: Pickering & Chatto.

Brennan, R.E., 1941. *Thomistic Psychology: A Philosophic Analysis of the Nature of Man.* New York: Macmillan.

Brewster, D., 1831. *The Life of Sir Isaac Newton.* London: John Murray.

Burnham, F.B., 1974. 'The More-Vaugham Controversy: the Revolt Against Philosophical Enthusiasm', *Journal of the History of Ideas*, 35: 33-49.

Buchdahl, G., 1969. *Metaphysics and the Philosophy of Science: The Classical Origins, Descartes to Kant.* Oxford: Oxford University Press.

Burtt, E.A., 1925. *The Metaphisycal Foundations of Modern Physical Science.* London: Paul, Trench, Trubner.

Butterfield, H., 1949. *The Origins of Modern Science, 1300-1800.* London: Bell.

Byrne, J.W., 1991. 'The Basis of the Natural Law in Locke's Philosophy', in R. Ashcraft (ed.), *John Locke Critical Assessments,* 4 vols. London: Routledge, ii, pp. 52-62.

Callon, M. and B. Latour, 1992. 'Don't throw the Baby Out with the Bath School! A Reply to Collins and Yearly', in A. Pickering (ed.), *Science as Practice and Culture.* Chicago: The University of Chicago Press.

Calvin, J., 1962. *Institutes of the Christian religion*, trans. Henry Beveridge, 2 vols. London: Clarke.

Cameron, E., 1992. *The European Reformation.* Oxford: Clarendon Press.

Cantor, G.N., 1983. *Optics after Newton.* Manchester: Manchester University Press.

Chalmers, A., 1993. 'The Lack of Excellency of Boyle's Mechanical Philosophy', *Studies in History and Philosophy of Science*, 24:541-64.

Chappell, V., 1994. 'Locke's theory of ideas', in V. Chappell (ed.), *The Cambridge Companion to Locke.* Cambridge: Cambridge University Press, pp. 26-55.

Charleton, W., 1654. *Physiologia Epicuro-Gassendo-Charletoniana: or, a Fabrick of Science Natural, upon the Hypothesis of Atoms*...London.

Chillingworth, W., 1638. *The Religion of Protestants a Safeway to Salvation,* London.

Chipman, R.A., 1954. 'An Unpublished Letter of Stephen Gray on Electrical Experiments, 1707-1708', *Isis,* 45: 33-50.

—, 1958. 'The Manuscript Letters of Stephen Gray, F.R.S. (1666/7-1736)', *Isis,* 49: 414-29.

Christensen, F., 1946. 'John Wilkins and the Royal Society's Reform of Prose Style', *Modern Language Quarterly,* 7: 179-87, 279-90.

Churchland, P.M. and C.A. Hooker (eds.), 1985. *Images of Science. Essays on Realism and Empiricism, with a Reply from Bas C. van Fraassen.* Chicago: The University of Chicago Press.

Clark, D.H. and L. Murdin, 1979. 'The Enigma of Stephen Gray Astronomer and Scientist (1666-1736)', *Vistas in Astronomy,* 23: 351-404.

Clericuzio, A., 1990. 'A Redifinition of Boyle's Chemistry and Corpuscular Philosophy', *Annals of Science,* 47: 561-589.

—, 1993. 'From van Helmont to Boyle. A Study of the Transmission of Helmontian Chemical and Medical Knowledge in Seventeenth-Century England', *British Journal for the History of Science,* 26: 303-334.

—, 1994. 'Carneades and the Chemists: A Study of *the Sceptical Chymist* and its Impact on Seventeenth-Century Chemistry.' In M. Hunter (ed.), *Robert Boyle Reconsidered.* Cambridge: Cambridge University Press, pp. 79-90.

Cohen, I.B., 1941. *Benjamin Franklin's Experiments.* Cambridge: Harvard University Press.

—, 1951. 'Guericke and Dufay', *Annals of Science,* 7: 207-9.

—, 1954. 'Neglected Sources for the Life of Stephen Gray', *Isis,* 45: 41-50.

—, 1956. *Franklin and Newton.* Philadelphia: American Philosophical Society.

—, 1958. 'Versions of Isaac Newton's First Published Paper', *Archive for History of Exact Sciences,* 11: 357-75.

— (ed.), 1958a. *Isaac Newton's Papers and Letters on Natural Philosophy and Related Documents.* Cambridge: Harvard University Press.

—, 1980. *The Newtonian Revolution.* Cambridge: Cambridge University Press.

— (ed.), 1990. *The Rise of Modern Science: The Merton Thesis.* New Brunswick: Rutgers University Press.

Collins, H.M., 1985. *Changing Order: Replication and Induction in Scientific Practice.* London: Sage.

Collins, H.M. and T. Pinch, 1998. *The Golem: What You Should Know About Science.* Cambridge: Cambridge University Press.

Conant. J.B. (ed), 1950. 'Robert Boyle's Experiments in Pneumatics', in *Harvard Case Histories in Experimental Science,* Case 1. Cambridge: Harvard University Press.

Cornford, F.M., 1997. *Plato's Cosmology. The* Timaeus *of Plato.* Indianapolis: Hackett Publishing Company.

Corson, D.W., 1968. 'Pierre Polinier, Francis Hauksbee, and Electroluminescence: A case of Simultaneous Discovery', *Isis*, 58: 402-13.

Costabel, P., 1982. 'La refraction de la lumière et la Dioptrique de Descartes', in *Démarches originales de Descartes savant*. Paris: Vrin.

Costello, W.T., 1958. *The Scholastic Curriculum of Early Seventeenth-Century Cambridge*. Cambridge: Harvard University Press.

Cottingham, J. (ed.), 1994. *Reason, Will, and Sensation. Studies in Descartes's Metaphysics.* Oxford: Clarendon Press.

Coudert, A., 1978. 'Some Theories of a Natural Language from Renaissance to the Seventeenth Century', *Studia Leibnitiana*, 7: 56-114.

Cragg, G.R., 1975. *Freedom and Authority. A Study of English Thought in Early Seventeenth Century.* Philadelphia: Westminster Press.

Crombie, A.C., 1953. *Robert Grosseteste and the Origins of Experimental Science.* Oxford: Clarendon Press.

—, 1967. 'The Mechanistic Hypothesis and the Scientific Study of Vision: some Optical Ideas as a Background to the Invention of the Microscope', in S. Bradbury and G.L.E. Turner, *Historical Aspects of Microscopy.* Cambridge: Heffer, pp. 3-112.

—, 1994. *Styles of Scientific Thinking in the European Tradition,* 2 vols. London: Duckworth.

Cudworth, R., 1678. *The True Intellectual System of the Universe, wherein all the Reason and Philosophy of Atheism is confuted, and its Impossibility Demonstrated.* London.

—, 1731. *A Treatise Concerning Eternal and Immutable Morality.* London.

Cunningham, A., 1988. 'Getting the Game Right: Some Plain Words on the Identity and Invention of Science', *Studies in History and Philosophy of Science*,19: 365-89.

—, 1991.'How the *Principia* Got its Name; or, Taking Natural Philosophy Seriously', *History of Science*, 29: 377-92.

Cunningham, A. and P. Williams, 1993. 'De-centring the 'Big Picture': *The Origins of Modern Science* and the Modern Origins of Science', *British Journal for the History of Science*, 26: 407-32.

Cuomo, S., 2000. 'Divide and Rule: Frontinus and Roman Land-Surveying', *Studies in History and Philosophy of Science*, 31: 189-202.

Daston, L., 1988. 'The Factual Sensibility', *Isis*, 79: 452-470.

—, 1991. 'Baconian Facts, Academic Civility, and the Prehistory of Objectivity', *Annals of Scholarship*, 8: 337-363.

Daston. L., and K. Park, 1998. *Wonders and the Order of Nature 1150-1750*, New York: Zone Books.

Davidson, H.M., 1983. *Blaise Pascal.* Boston: Twayne Publishers.

Dear, P., 1985. '*Totius in verba.* Rhetoric and Authority in the Early Royal Society', *Isis*, 76: 145-61.

—, 1990. 'Miracles, Experiments, and the Ordinary Course of Nature', *Isis*, 81: 663-83.

—, 1995. *Discipline and Experience. The Mathematical Way in the Scientific Revolution.* Chicago: The University of Chicago Press.

— (ed.), 1991. *The Literary Structure of Scientific Arguments: Historical Studies.* Philadelphia: University of Pennsylvania Press.

—, 1998. 'Method and the Study of Nature', in D. Garber and M. Ayers (eds.), *The Cambridge History of Seventeenth-Century Philosophy.* Cambridge: Cambridge University Press, pp. 147-77.

Debus, A.G., 1962. 'Solution Analyses Prior to Robert Boyle', *Chymia*, 8: 41-60.

—, 1974. 'The Chemical Philosophers: Chemical Medicine from Paracelsus to van Helmont', *History of Science*, 12: 235-259.

—, 1978. *Man and Nature in the Renaissance.* Cambridge: Cambridge University Press.

Derham, W. (ed.), 1726. *Miscellanea Curosa. Containing a Collection of some of the Principal Phaenomena in Nature*, 3 vols., 3d ed. London.

Desaguliers, J.T., 'An Account of Some Experiments of Light and Colours...', *Philosophical Transactions*, 29: 433-52.

—, 1717. *Physico-Mechanical Lectures, or, an Account of what is Explain'd and Demonstrated in the Course of Mechanical and Experimental Philosophy.* London.

—, 1722. 'An Account of an Optical Experiments made before the Royal society', *Philosophical Transactions*, 32: 206-8.

—, 1728. 'Optical Experiments...', *Philosophical Transactions*, 35: 596-629.

—, 1734. *A Course of Experimental Philosophy*, 2 vols. London.

—, 1739-40. 'Some Thoughts and Conjectures Concerning the Cause of Electricity', *Philosophical Transactions*, 41: 175-85.

—, 1739-40a. 'Some Thought and Experiments Concerning Electricity', *Philosophical Transactions*, 41: 186-93.

—, 1744. *A Course of Experimental philosophy*, ii, London.

—, 1763. *A Course of Experimental Philosophy*, 2 vols. London.

Descartes, R., 1677. *Specimina Philosophiae.* Amstelodami: Apud Danielem Elsevirium.

—, 1971. *Descartes Philosophical Writings,* ed. Elizabeth Anscombe and Peter T. Geach. Indianapolis: The Library of Liberal Arts.

—, 1982-1989. *Oeuvres de Descartes*, ed. C. Adam and P. Tannery. Paris: Vrin.

—, 1983. *Principles of Philosophy*, trans. V. R. Miller and R. P. Miller. Dordrecht: D. Reidel.

—, 1985. *The Philosophical Writings of Descartes,* ed. and trans. J. Cottingham, R. Stoothoff, and D. Murdoch, 2 vols. Cambridge: Cambridge University Press.

—, 1991. *The Philosophical Writings of Descartes*, ed. and trans. J. Cottingham, R. Stoothoff, D. Murdoch, and A. Kenny, 3rd vol. Cambridge: Cambridge University Press.

—, 1998. *The World and Other Writings*, trans. and ed. S. Gaukroger. Cambridge: Cambridge University Press.

Dickens, A.G., 1964. *The English Reformation.* New York: Schocken Books.

Dobbs, B.J.T., 1990. 'Newton as Alchemist and Theologian', in N.J.W. Thrower, *Standing on the Shoulders of Giants. A Longer View of Newton and Halley.* Berkeley: University of California Press, pp. 128-40.

—, 1991. *The Janus Face of Genius. The Role of Alchemy in Newton's Thought.* Cambridge: Cambridge University Press.

—, 2000. 'Newton as Final Cause and First Mover', in M.J. Osler (ed.), *Rethinking the Scientific Revolution.* Cambridge: Cambridge University Press, pp. 25-39.

Dobbs, B.J.T., and M.C. Jacob, 1995. *Newton and the Culture of Newtonianism.* Atlantic Highlands: Humanities Press.

Drake, S. (ed.), 1957. *Discoveries and Opinions of Galileo.* New York: Doubleday Anchor.

Driscoll, E.A., 1972. 'The influence of Gassendi on Locke's hedonism', *International Philosophical Quarterly*, 12: 87-110.

Dufay, C., 1733. 'Premier Mémoire sur l'électricité…', 'Second Mémoire…', 'Troisiéme Mémoire…', *Mémoires de l'Académie Royale des Sciences*, 23-35, 73-84, 233-53.

—, 1733-4. 'A letter from Mons. Du Fay..concerning Electricity', *Philosophical Transactions*, 38: 258-66.

Dunn, J., 1969. *The Political Thought of John Locke.* Cambridge: Cambridge University Press.

—, 1986. 'Trust in the Politics of John Locke', in *Rethinking Modern Political Theory. Essays 1979-83.* Cambridge: Cambridge University Press.

—, 1992. *Locke,* in *The British Empiricists.* Oxford: Oxford University Press, pp. 1-90.

Easton, P.A. (ed.), 1997. *Logic and the Workings of the Mind: The Logic of Ideas and Faculty Psychology in Early Modern Philosophy.* Atascadero: Ridgeview Publishing.

Eastwood, B.S., 1966. 'Robert Grosseteste's Theory of the Rainbow', *Archives internationales d'histoire des sciences*, 19: 313-32.

—, 1984. 'Descartes on Refraction: Scientific versus Rhetorical Method', *Isis,* 75: 481-502.

Eisenstein, E.L., 1980. *The Printing Press as an Agent of Change: Communications and Cultural Transformations in Early-Modern Europe.* Cambridge: Cambridge University Press.

Elzinga, A., 1980. 'Christian Huygens' Theory of Research', *Janus*, 67: 281-300.

Espinasse, M., 1958. 'The Decline and Fall of Restoration Science', *Past and Present*, 14: 71-89.

Evelyn, J., 1955. *The Diary of John Evelyn*, ed. E.S. de Beer, 6 vols. Oxford: Clarendon Press.

Ezrahi, Y., 1990. *The Descent of Icarus: Science and the Transformation of Contemporary Democracy*. Cambridge: Harvard University Press.

Feingold, M., 1984. *The Mathematicians' Apprenticeship: Science, Universities and Society in England, 1560-1640*. Cambridge: Cambridge University Press.

—, 1990. 'Isaac Barrow's Library', in M. Feingold (ed.), *Before Newton. The Life and Times of Isaac Barrow*. Cambridge: Cambridge University Press, pp. 333-372.

—, 2001. 'Mathematicians and Naturalists: Sir Isaac Newton and the Royal Society', in J.Z. Buchwald and I.B. Cohen (eds.), *Isaac Newton's Natural Philosophy*. Cambridge: Cambridge University Press, pp. 77-102.

Feldhay, R., 1995. *Galileo and the Church. Political Inquisition or Critical Dialogue?* Cambridge: Cambridge University Press.

Feyerabend, P.K., 1970. 'Classical Empiricism', in R.E. Butts and J.W. Davies (eds.), *The Methodological Heritage of Newton*. Oxford: Blackwell, pp. 150-70.

Findlen, P., 1994. *Possessing Nature: Museums, Collecting and Scientific Culture in Early Modern Italy*. Berkeley: University of California Press.

Finocchiaro, M.A., 1989. *The Galileo Affair. A Documentary History*. Berkeley: The University of California Press.

Fisch, H., 1952. 'The Puritans and the Reform of Prose-Style', *Journal of English Literary History*, 19: 229-48.

—, 1953. 'The Scientist as Priest: A Note on Robert Boyle's Natural Theology', *Isis*, 44: 252-265.

Fleck, L., 1979. *Genesis and Development of a Scientific Fact*, ed. T.J. Trenn and R.K. Merton, trans. F. Bradley and T.J. Trenn. Chicago: The University of Chicago Press.

Force, 1990. 'Newton's God of Dominion: The Unity of Newton's Theological, Scientific, and Political Thought', in J.E. Force and R.H. Popkin, *Essays on the Context, Nature, and Influence of Isaac Newton's theology*. Dordrecht: Kluwer Academic Publishers, pp. 75-102.

—, 1990. 'Newton's 'Sleeping Argument and the Newtonian synthesis of Science and Religion', in N.J.W. Thrower, *Standing on the Shoulders of Giants. A Longer View of Newton and Halley*. Berkeley: University of California Press, pp. 109-27.

—, 1999. ''Children of the Resurrection' and 'Children of Dust': Confronting Mortality and Immortality with Newton and Hume', in J.E. Force and

D.S. Katz (eds.), *Everything Connects: In Conference with Richard H. Popkin. Essays in his Honor*. Boston: Brill, pp. 115-42.

Force, J.E., and R.H. Popkin (eds.), 1999. *Newton and Religion. Context, Nature, and Influence*. Dordrecht: Kluwer Academic Publishers.

Fowler, E., 1670. *The Principles and Practices of Certain Moderate Divines of the Church of England, (greatly mis-understood) Truly Represented and Defended.* London.

Frank Jr., R.G., 1980. *Harvey and the Oxford Physiologists. Scientific Ideas and Social Interaction.* Berkeley: University of California Press.

Fraser Mitchell, W., 1962. *English Pulpit Oratory, from Andrewes to Tillotson: A Study of Its Literary Aspects.* New York: Russell & Russell.

Freudenthal, Gad, 1981. 'Early Electricity Between Chemistry and Physics: The Simultaneous Itineraries of Francis Hauksbee, Samuel Wall and Pierre Poliniere', *Historical Studies in the Physical Sciences*, 9: 203-29.

—, 1983. 'Theory of Matter and Cosmology in William Gilbert's *De magnete*', *Isis*, 74: 22-37.

—, 1987. 'Joseph Ben-David's Sociology of Scientific Knowledge', *Minerva*, 25: 135-49.

Funkenstein, A., 1986. *Theology and the Scientific Imagination from the Middle Ages to the Seventeenth Century.* Princeton: Princeton University Press.

Gabbey, A., 1993. 'Between *ars* and p*hilosophia naturalis*: Reflections on the Historiography of Early Modern Mechanics', in J. Field and F. James (eds.), *Renaissance and Revolution.* Cambridge: Cambridge University Press, pp. 133-145.

Galison, P.L., 1987. *How Experiments End.* Chicago: The University of Chicago Press.

Garber, D., 1992. *Descartes' Metaphysical Physics*. Chicago: The University of Chicago Press.

—, 1993. 'Descartes and Experiment in the *Discourse* and *Essays*', in S. Voss (ed.), *Essays on the Philosophy and Science of René Descartes.* Oxford: Clarendon Press, pp. 288-323.

—, 2001. *Descartes Embodied. Reading Cartesian Philosophy through Cartesian Science.* Cambridge: Cambridge University Press.

Gascoigne, J., 1985. 'The Universities and The Scientific Revolution: The Case of Newton and Restoration Cambridge', *History of Science*, 23: 391-434.

Gassendi, P., 1981. *Pierre Gassendi's Institutio Logica 1658*, ed. and trans. H. Jones. Assen: Van Gorcum.

Gaukroger, S., 1978. *Explanatory Structures. A Study of Concepts of Explanation in Early Physics and Philosophy.* New Jersey: Humanities Press.

—, 1994. 'The Sources of Descartes's Procedure of Deductive Demonstration in Metaphysics and Natural Philosophy', in J. Cottingham (ed.), *Reason, Will, and Sensation. Studies in Descartes's Metaphysics.* Oxford: Clarendon Press.

—, 2001. *Francis Bacon and the Transformation of Early-Modern Philosophy.* Cambridge: Cambridge University Press.

Geertz, C., 1973. 'Ideology As a Cultural System', in *The Interpretation of Cultures.* New York: Basic Books, pp. 193-233.

Gellner, E., 1970. 'Concepts and Society', in B. Wilson (ed.), *Rationality.* Oxford: Blackwell.

—, 1979. 'Notes Towards a Theory of Ideology', in *Spectacles & Predicaments: Essays in Social Theory.* Cambridge: Cambridge University Press.

Geree, J., 1646. *The Character of an Old English Puritan or Non-conformist.* London.

Gilbert, N.W., 1960. *Renaissance Concepts of Method.* New York: Columbia University Press.

Gilbert, W., 1941. *De Magnete,* trans. P.F. Mottelay. Ann Arbor: Edwards Brothers.

Gillispie, C.C., 1960. *The Edge of Objectivity.* Princeton: Princeton University Press.

Gilson, E., 1991. *The Spirit of Mediaeval Philosophy (Gifford Lectures 1931-1932),* trans. A.H.C. Downes. Notre Dame: University of Notre Dame Press.

Glanvill, J., 1668. *Plus Ultra: or, the progress and advancement of knowledge since the days of Aristotle.* London.

—, 1676. *Essays on Several Important subjects in Philosophy and Religion,* London.

—, 1678. *A Seasonable Defense of Preaching.* London.

Golinski, J.V., 1989. 'A Noble Spectacle: Phosphorus and the Public Cultures of Science in the Early Royal Society', *Isis,* 80: 11-39.

—, 1998. *Making Natural Knowledge. Constructivism and the History of Science.* Cambridge: Cambridge University Press.

Grafton, A., 1992. *New World, Ancient Texts: The Power of Tradition and the Shock of Discovery.* Cambridge: Belknap Press of Harvard University Press.

Grant, E., 1981. *Much Ado about Nothing: Theories of Space and Vacuum from the Middle Ages to the Scientific Revolution.* Cambridge: Cambridge University Press.

Gray, S., 1720. 'An Account of some new Electrical Experiments', *Philosophical Transactions,* 31: 140-8.

—, 1731. 'A letter to Cromwell Mortimer, M.D. Secretary of the Royal Society containing several Experiments concerning Electricity', *Philosophical Transactions,* 37: 18-44.

—, 1732. 'The Electricity of Water', *Philosophical Transactions,* 37: 227-30.

—, 1732a. 'Farther Account of his Experiments concerning Electricity', *Philosophical Transactions,* 37: 285-91.

—, 1735. 'Experiments and Observations upon the Light...', *Philosophical Transactions*, 39: 16-24.

—, 1736. 'Some Experiments relating to Electricity', *Philosophical Transactions*, 39: 166-70.

—, 1736a. 'His last Letter...', *Philosophical Transactions*, 39: 220.

Greenblatt, S., 1991. *Marvelous Possessions: The Wonder of the New World.* Chicago: The University of Chicago Press.

Greene, R.A., 1981. 'Whichcote, Wilkins, 'Ingenuity', and the Reasonableness of Christianity', *Journal of the History of Ideas*, 42: 227-52.

Guerlac, H., 1964. 'Sir Isaac and the Ingenious Mr. Hauksbee', in R. Taton and I.B. Cohen (eds.), *Mélanges Alexander Koyré.* 2 vols., Paris: Hermann, i, pp. 228-53.

—, 1977. 'Newton's Optical Aether: His Draft of a Proposed Addition to His *Optics*', in *Essays and Paper in the History of Modern Science,* Baltimore: Johns Hopkins University Press, pp. 120-130.

—, 1981. *Newton on the Continent*. Ithaca: Cornell University Press.

—, 1983. 'The Dating of Newton's Early Optical Experiments', *Isis*, 74: 74-80.

Guicciardini, N., 1999. *Reading the* Principia*: The Debate on Newton's Mathematical Methods for Natural Philosophy from 1687 to 1736.* Cambridge: Cambridge University Press.

Guyer, P., 1994. 'Locke's Philosophy of Language', in V. Chappell (ed.), *The Cambridge Companion to Locke.* Cambridge: Cambridge University Press, pp. 115-45.

Hacking, I., 1975. *The Emergence of Probability: A Philosophical Study of Early Ideas about Probability, Induction and Statistical Inference.* Cambridge: Cambridge University Press.

—, 1983. *Representing and Intervening.* Cambridge: Cambridge University Press.

Hackmann, W.D., 1978. *Electricity from Glass: The History of the Frictional Electrical Machine 1600-1850.* Alphen aan den Rijn: Sijhoff & Noordhiff.

—, 1979. 'The Relationship Between Concept and Instrument Design in Eighteenth-Century Experimental Science', *Annals of Science*, 36: 205-24.

Haddon, R.W., 1994. *On the Shoulders of Merchants: Exchange and the Mathematical conception of Nature in Early Modern Europe.* Albany: State University of New York Press.

Hall, A.R., 1947-9. 'Sir Isaac Newton's Note-Book', *Cambridge Historical Journal*, 9: 239-50.

—, 1955. 'Further Optical Experiments of Isaac Newton', *Annals of Science*, 11: 27-43.

—, 1996. *Isaac Newton. Adventurer in Thought.* Cambridge.

Hall, A.R. and M. Boas Hall, 1962. *Unpublished Scientific Papers of Isaac Newton.* Cambridge: Cambridge University Press.

—, 1962a. 'Why Blame Oldenburg?', *Isis*, 53: 482-91.

——, 1964. 'Philosophy and Natural Philosophy: Boyle and Spinoza', in R. Taton and I.B. Cohen (eds.), *Mélanges Alexander Koyré.* 2 vols. Paris: Hermann, ii, pp. 241-56.

Hankins, T.L., and R.J. Silverman, 1995. *Instruments and the Imagination.* Princeton: Princeton University Press.

Harris, I., 1994. *The Mind of John Locke. A Study of Political Theory in its Intellectual Setting*. Cambridge: Cambridge University Press.

Harrison, J., 1978. *The Library of Isaac Newton.* Cambridge: Cambridge University Press.

Harrison J. and P. Laslett, 1965. *The Library of John Locke*. Oxford: Oxford University Press.

Hatfield, G., 1997. 'The Workings of the Intellect: Mind and Psychology', in P.A. Easton, *Logic and the Workings of the Mind: The Logic of Ideas and Faculty Psychology in Early Modern Philosophy.* Atascadero: Ridgeview Publishing.

——, 1998. 'The Cognitive Faculties', in D. Garber and M. Ayers (eds.), *The Cambridge History of Seventeenth-Century Philosophy.* Cambridge: Cambridge University Press, pp. 953-1002.

Hauksbee, F., 1705. 'Several Experiments on the Mercurial Phosphorous made before the Royal Society, at Gresham-College', *Philosophical Transactions*, 24: 2129-36.

——, 1705a. 'Several Experiments on Attrition of Bodies in Vacuo', *Philosophical Transactions*, 24: 2165-75.

——, 1706. 'An account of an Experiment..touching the Production of a Considerable Light.', *Philosophical Transactions*, 25: 2277-82.

——, 1706a. 'An Account of an Experiment..touching the Extraordinary Elistricity of Glass..', *Philosophical Transactions*, 25: 2327-35.

——, 1707. 'Several Experiments shewing the strange Effects of the Effluvia of Glass', *Philosophical Transactions*, 25: 2372-77.

——, 1719. *Physico-Mechanical Experiments on Various Subjects,* 2nd ed. London.

Hawes, J.L., 1968. 'Newton and the Electrical Attraction Unexcited', *Annals of Science*, 24: 121-30.

——, 1971. 'Newton's Two Electricities', *Annals of Science*, 27: 95-101.

Heathcote, N.H. de W., 1950. 'Guericke's Sulphur Globe', *Annals of Science*, 6: 293-305.

——, 1967. 'The Early Meaning of Electricity: Some Pseudodoxia Epidemica', *Annals of Science*, 23: 261-76.

Heilbron, J.L., 1979. *Electricity in the 17th and 18th Centuries: A Study of Early Modern Physics.* Berkeley: University of California Press.

——, 1980. 'Experimental Natural Philosophy', in G.S. Rousseau and R. Porter (eds.), *The Ferment of Knowledge: Studies in the Historiography of Eighteenth-Century science*, Cambridge: Cambridge University Press.

—, 1983. *Physics at the Royal Society During Newton's Presidency.* Los Angeles: University of California Press.

Heimann P.M. and J.E. McGuire, 1971. 'Newtonian Forces and Lockean Powers: Concepts of Matter in Eighteenth-Century Thought', *Historical Studies in the Physical Sciences*, 3: 233-306.

Henry, J., 1982. 'Thomas Harriot and Atomism: A Reappraisal', *History of Science*, 20: 267-96.

—, 1986. 'Occult Qualities and the Experimental Philosophy: Active Principles in pre-Newtonian Matter Theory', *History of Science,* 24: 335-81.

—, 1990. 'Henry More versus Robert Boyle: The Spirit of Nature and the Nature of Providence', in S. Hutton (ed.), *Henry More (1614-87): Tercentenary Studies*, Dordrecht: Kluwer Academic Publishers, pp. 55-76.

—, 1992. 'The Scientific Revolution in England', in R. Porter and M. Teich,
The Scientific Revolution in National context. Cambridge: Cambridge University Press, pp. 178-209.

—, 1994. ''Pray Do Not Ascribe that Notion to Me: God and Newton's Gravity', in J.E. Force and R.H. Popkin (eds.), *The Book of Nature and Scripture: Recent Essays on Natural Philosophy, Theology, and Biblical Criticism in the Netherlands of Spinoza's Time and the British Isles of Newton's Time.* Dordrecht: Kluwer Academic Publishers, pp. 123-47.

—, 1994a. 'Boyle and Cosmical Qualities', in M. Hunter (ed.), *Robert Boyle Reconsidered.* Cambridge: Cambridge University Press, pp. 119-38.

—, 1997. *The Scientific Revolution and the Origins of Modern Science.* London: Macmillan Press.

Hesse, M., 1963. *Models and Analogies in Science.* South Bend: University of Notre Dame Press.

—, 1964. 'Hooke's Development of Bacon's Method', *Proceedings of the Tenth International Congress of History of Science*, 1: 265-73.

Heyd, M., 1981. 'The Reaction to Enthusiasm in the Seventeenth Century: Towards an Integrative Approach', *Journal of Modern History*, 53: 258-80.

—, 1988. 'The emergence of modern science as an autonomous world of knowledge in the Protestant tradition of the seventeenth century,' in *Cultural Traditions and Worlds of Knowledge: Explorations in the Sociology of Knowledge*, 7:165-179.

Hill, C., 1969. ''Reason' and 'Reasonableness' in Seventeenth-Century England', *British Journal of Sociology,* 20: 235-52.

Hintikka, J. and U. Remes, 1974. *The Method of Analysis: Its Geometrical Origin and Its General Significance.* Dordrecht: D. Reidel.

Hirschman, A.O., 1977. *The Passions and the Interests. Political Arguments for Capitalism before Its Triumph.* Princeton: Princeton University Press.

Hiscock, W.G. (ed.), 1937. *David Gregory, Isaac Newton and their Circle. Extracts from David Gregory's Memoranda 1677-1708.* Oxford.

Hobbes, T., 1840. *The English Works of Thomas Hobbes of Malmesbury*, ed. W. Molesworth, 11 vols. London.

—, 1975. *Leviathan*, ed. C.B. Macpherson. Harmondsworth: Pelican.

—, 1985. *A Physical Dialogue*, trans. S. Schaffer, in S. Shapin and S. Schaffer, *Leviathan and the air-pump. Hobbes, Boyle, and the experimental life*. Princeton: Princeton University Press, pp. 345-391.

Hobson, M., 1982. *The Object of Art: The Theory of Illusion in 18th Century France*. Cambridge: Cambridge University Press.

Holton, G., 1975. *Thematic Origins of Scientific Thought: Kepler to Einstein*. Cambridge: Harvard University Press.

Home, R.W., 1967/8. 'Francis Hauksbee's Theory of Electricity', *Archive for History of Exact Sciences*, 4: 203-17.

—, 1981. *The Effluvial Theory of Electricity*. New York.

—, 1982. 'Newton on Electricity and the Aether', in Z. Bechler (ed.), *Contemporary Newtonian Research*. Dordrecht: D. Reidel, pp. 191-213.

—, 1985. 'Force, Electricity, and the Powers of Living Matter in Newton's Mature Philosophy of Nature', in M.J. Osler and P.L. Farber, *Religion, Science, and Worldview. Essays in Honor of Richard S. Westfall*. Cambridge: Cambridge University Press.

Hooke, R., 1961 . *Micrographia: Or some physiological descriptions of minute bodies made by magnifying glasses, with observations and inquiries thereupon*. New York: Dover Publications.

—, 1969. *The Posthumous Works of Robert Hooke*, ed. R. Waller. The Sources of Science, no.73.

Hooykass, R., 1990. 'Science and Reformation', in I.B. Cohen (ed.), *Puritanism and the rise of modern science: The Merton Thesis*. New Brunswick: Rutgers University Press, pp. 189-99.

Houghton, W.E., 1942. 'The English Virtuoso in the Seventeenth Century', *Journal of the History of Ideas*, 3: 51-73, 190-219.

Hunter, M., 1981. *Science and Society in Restoration England*, Cambridge.

—, 1981-2. 'Early Problems in Professionalizing Scientific Research: Nehemiah Grew (1641-1712) and the Royal Society, with an Unpublished Letter to Henry Oldenburg', *Notes and Records of the Royal Society*, 36: 188-209.

—, 1982. *The Royal Society and its Fellows 1660-1700: The Morphology of an Early Scientific Institution*, British Society for the History of Science Monographs iv.

—, 1983-4. 'A 'College' for the Royal Society: The Abortive Plan', *Notes and Records of the Royal Society*, 38: 59-186.

—, 1989. 'Latitudinarianism and the 'Ideology' of the Early Royal Society: Thomas Sprat's *History of the Royal Society* (1667) Reconsidered', in *Establishing the New Science: The Experience of the Early Royal Society*. Woodbridge: Boydell Press.

—, 1995. 'How Boyle became a scientist', *History of Science*, 33: 59-103.

—, 1999. 'Robert Boyle (1627-91): A Suitable Case for Treatment?', *British Journal for the History of Science*, 32: 261-75.

Hunter, M. and P. B. Wood, 1986. 'Towards Solomon's House: Rival Strategies for Reforming the Early Royal Society', *History of Science*, 24: 49-108.

Hutchison, K., 1982. 'What Happened to Occult Qualities in the Scientific Revolution?', *Isis*, 73: 233-53.

—, 1983. 'Supernaturalism and the Mechanical Philosophy', *History of Science*, 21: 297-333.

Hutton, S., 1994. 'More, Newton, and the Language of Biblical Prophecy', in J.E. Force and R.H. Popkin (eds.), *The Book of Nature and Scripture: Recent Essays on Natural Philosophy, Theology, and Biblical Criticism in the Netherlands of Spinoza's Time and the British Isles of Newton's Time.* Dordrecht: Kluwer Academic Publishers, pp. 39-53.

Huygens, C., 1912. *Treatise on Light*, trans. S.P. Thompson. Chicago.

Iliffe, R., 1992. ''In the Warehouse': Privacy, Property and Priority in the Early Royal Society', *History of Science*, 30: 29-68.

—, 1994. ''Making a Shew': Apocalyptic Hermeneutics and the Sociology of Christian Idolatry in the Work of Isaac Newton and Henry More', in J.E. Force and R.H. Popkin (eds.), *The Book of Nature and Scripture: Recent Essays on Natural Philosophy, Theology, and Biblical Criticism in the Netherlands of Spinoza's Time and the British Isles of Newton's Time.* Dordrecht: Kluwer Academic Publishers, pp. 55-88.

—, 1999. 'Those 'Whose Business It Is To Cavill': Newton's Anti-Catholicism', in J.E. Force and R.H. Popkin (eds.), *Newton and Religion. Context, Nature, and Influence.* Dordrecht: Kluwer Academic Publishers, pp. 97-119.

Impey, O., and A. MacGregor (eds.), 1985. *The Origins of Museums: The Cabinet of Curiosities in Sixteenth and Seventeenth-Century Europe.* Oxford: Clarendon Press.

Jacob, J.R., 1977. *Robert Boyle and the English Revolution: A Study in Social and Intellectual Change.* New York: Burt Franklin.

Jacob, J.R. and M.C. Jacob, 1980. 'The Anglican Origins of Modern Science: the Metaphysical Foundations of the Whig Constitution', *Isis,* 71: 251-67.

Jacob, M.C., 1976. *The Newtonians and the English Revolution, 1689-1720.* Ithaca: Cornell University Press.

Jacobs, S., 1994. 'Laws of Nature, Corpuscles, and Concourse: Non-occasionalist Tendencies in the Natural Philosophy of Robert Boyle, *Journal of Philosophical Research*, 19:373-93.

Jardine, L., 1974. *Francis Bacon. Discovery and the Art of Discourse.* Cambridge: Cambridge University Press.

Jardine, N., 1988. 'Epistemology of the Sciences', in C.B. Schmitt *et al.* (eds.), *The Cambridge History of Renaissance Philosophy.* Cambridge: Cambridge University Press, pp. 685-711.

—, 1991. *The Scenes of Inquiry: On the Reality of Questions in the Sciences.* Oxford: Clarendon Press.

Jolley, N., 1998. 'The Relation between Theology and Philosophy', in D. Garber and M. Ayers (eds.), *The Cambridge History of Seventeenth-Century Philosophy.* Cambridge: Cambridge University Press, pp. 363-92.

Jones, M.L., 2001. 'Writing and *Sentiment*: Blaise Pascal, the Vacuum, and the *Pensées*', 32: 139-81.

—, 2001a. 'Descartes's Geometry as Spiritual Exercise', *Critical Inquiry*, 28: 40-71.

Jones, R.S., 1930. 'Science and English Prose Style in the Third Quarter of the Seventeenth Century', *Publications of Modern Language Association*, 45.2: 977-1009.

—, 1932. 'Science and Language in England of the Mid-Seventeenth Century', *Journal of English and Germanic Philology*, 31: 315-31.

—, 1965. 'The Attack on Pulpit Eloquence in the Restoration: an Episode in the Development of the Neo-Classical Standard for Prose', in R.S. Jones (ed.), *The Seventeenth Century. Studies in the History of English Thought and Literature from Bacon to Pope.* Stanford: Stanford University Press, pp. 111-42.

Jordan, W., 1992. *Ancient Concepts of Philosophy.* London: Routledge.

Kay, C.D., 1981. *William Gilbert's Renaissance Philosophy of the Magnet.* Doctoral dissertation, University of Pittsburgh.

Keill, J., 1720. *An Introduction to Natural Philosophy.* London.

Klaaren, E.M., 1977. *Religious Origins of Modern Science: Belief in Creation in Seventeenth-Century Thought.* Grand Rapids: Eerdmans.

Knorr-Cetina, K.D., and M.J. Mulkay (eds.), 1983. *Science Observed: Perspectives on the Social Study of Science.* London: Sage.

Knowlson, J., 1975. *Universal Language Schemes in England and France 1600-1700.* Toronto: University of Toronto Press.

Koyré, A.,1965. *Newtonian Studies.* London: Chapman & Hall.

Koyré, A. and I.B. Cohen, 1960. 'Newton's Electric and Elastic Spirit', *Isis*, 51: 337.

Kripke, S., 1982. *Wittgenstein on Rules and Private Language.* Cambridge: Harvard University Press.

Kroll, R.W.F., 1984 'The Question of Locke's Relation to Gassendi', *Journal of the History of Ideas*, 45: 339-59.

Kuhn, T., 1952. 'Robert Boyle and Structural Chemistry in the Seventeenth Century', *Isis*, 43: 12-36.

—, 1958. 'Newton's Optical Papers', in I.B. Cohen (ed.), *Isaac Newton's Papers and Letters on Natural Philosophy.* Cambridge: Cambridge University Press.

—, 1962. *The Structure of Scientific Revolutions.* Chicago: The University of Chicago Press.

—, 1963. 'The Function of Dogma in Scientific Research', in A.C. Crombie (ed.), *Scientific Change*, New York: Basic Books, pp. 347-69, 386-95.

—, 1977. *The Essential Tension.* Chicago: The University of Chicago Press.

Laird, W.L., 1997. 'Galileo and the Mixed Sciences', in D.A. Di Liscia, E. Kessler, C. Methuen, *Method and Order in Renaissance Philosophy of Nature. The Aristotle Commentary Tradition.* Aldershot: Ashgate.

Lakatos, I., 1970. 'Falsification and the Methodology of Scientific Research Programmes', in I. Lakatos and A. Musgrave (eds.), *Criticism and the Growth of Knowledge.* Cambridge: Cambridge University Press, pp. 91-195.

Latour, B., 1987 *Science in Action.* Milton Keynes: Open University Press.

—, 1999. 'For David Bloor...and Beyond: A reply to David Bloor's 'Anti-Latour'', *Studies in History and Philosophy of Science*, 30:113-29.

Latour, B. and S. Woolgar, 1979. *Laboratory Life. The Social Construction of Scientific Facts.* Beverly Hills: Sage.

Laudan, L., 1966. 'The Clock Metaphor and Probabilism: The Impact of Descartes on English Methodological Thought, 1650-65', *Annals of Science*, 22: 73-104.

Law, J., 1993. *Modernity, Myth, and Materialism.* Oxford: Blackwell.

Law, J., M. Callon, and A. Rip (eds.), 1986. *Mapping the Dynamics of Science and Technology: Sociology of Science in the Real World.* Houndmills: Macmillan.

Law, J.D., 1993. *The Rhetoric of Empiricism: Language and Perception from Locke to I.A. Richards.* Ithaca: Cornell University Press.

Laymon, R., 1978. 'Newton's Advertised Precision and His Refutation of the Received Laws of Refraction', in P.K. Machamer and R.G. Turnbull (eds.), *Studies in Perception: Interrelations in the History of Philosophy and Science.* Columbus: Ohio State University Press, pp. 231-258.

Lear, J., 1988. *Aristotle: The Desire to Understand.* Cambridge: Cambridge University Press.

Lennox, J., 1983. 'Robert Boyle's Defense of Teleological Inference in Experimental Science', *Isis*, 74: 38-52.

Lenoir, T., 1997. *Instituting Science: The Cultural Production of Scientific Disciplines.* Stanford: Stanford University Press.

Lindberg, D.C., 1968. 'The Cause of Refraction in Medieval Optics', *British Journal for the History of Science*, 4: 23-38.

—, 1976. *Theories of Vision from al-Kindi to Kepler.* Chicago: The University of Chicago Press.

—, 1983. 'The Science of Optics', in D.C. Lindberg (ed.), *Science in the Middle Ages*. Chicago: The University of Chicago Press, pp. 338-68.

—, 1990. 'Conceptions of the Scientific Revolution from Bacon to Butterfield: A Preliminary Sketch', in D.C. Lindberg and R.S. Westman (eds.), *Reappraisals of the Scientific Revolution*. Cambridge University Press, pp. 1-26.

Lindberg, D.C. and R.S. Westman (eds.), 1990. *Reappraisals of the Scientific Revolution*. Cambridge: Cambridge University Press.

Lloyd, G.E.R., 1979. *Magic, Reason and Experience. Studies in the Origin and Development of Greek Science*. Cambridge: Cambridge University Press.

Locke, J., 1849. *Locke's Essays. An Essay Concerning Human Understanding* and *A Treatise on the Conduct of the Understanding*. Philadelphia: Kay & Troutman.

—, 1884. *The Life and Letters of John Locke*, ed. Lord King. New York: Burt Franklin.

—, 1958. *Essays on the Law of Nature*, trans. and ed. W. von Leyden. Oxford: The Clarendon Press.

—, 1965. *Two Treatises of Government,* ed. with introduction P. Laslett. Cambridge: Cambridge University Press.

—, 1975. *An Essay concerning Human Understanding*, ed. with introduction P.H. Nidditch. Oxford: Clarendon.

—, 1976. *The Correspondence of John Locke*, ed. E.S. de Beer, 8 vols. Oxford: The Clarendon Press.

—, 1989. *The Reasonableness of Christianity.* Washington: Regnery Gateway.

—, 1989a. *Some Thought concerning Education,* ed. J.W. and J.S. Yolton. Oxford: Clarendon Press.

—, 1990. *Drafts for the Essay Concerning Human Understanding, and Other Philosophical Writings*, ed. P.H. Nidditch and G.A.L. Rogers. Oxford: Clarendon Press.

—, 1993. *Of the Conduct of the Understanding*, ed. with introduction J. Yolton. Bristol: Thoemmes Press.

—, 1999. *Questions Concerning the Law of Nature*, ed. and trans. with introduction R. Horwitz, J. Strauss Clay, D. Clay. Ithaca: Cornell University Press.

—, 1999a. *The Reasonableness of Christianity As delivered in the Scriptures*, ed. with introduction J.C. Higgins-Biddle. Oxford: Clarendon Press.

Lohne, J., 1959. 'Thomas Hariott (1560-1621): The Tycho Brahe of Optics', *Centaurus*, 6: 113-21.

—, 1965. 'Isaac Newton: The Rise of a Scientist 1661-1671', *Notes and Records of the Royal Society*, 20: 125-39.

—, 1968. 'Experimentum Crucis', *Notes and Records of the Royal Society*, 23: 169-99.

Losonsky, M., 1994. 'Locke on Meaning and Signification', in G.A.J. Rogers (ed.), *Locke's Philosophy. Content and Context.* Oxford: Clarendon Press, pp. 123-141.

Lowe, E.J., 1995. *Locke on Human Understanding.* London: Routledge.

Lynch, M., 1984. *Art and Artifact in Laboratory Science: A Study of Shop Work and Shop Talk in a Research Laboratory.* London: Routledge & Kegan Paul.

—, 1992. 'Extending Wittgenstein: The Pivotal Move from Epistemology to the Sociology of Science', in A. Pickering (ed.), *Science as Practice and Culture.* Chicago: The University of Chicago Press.

Mach, E., 1926. *The Principles of Physical Optics.* London: Dover Publications.

MacIntyre, A., 1989. 'The Relationship of Philosophy to History: Postscript to the Second Edition of *After Virtue*', in Kenneth Baynes, James Bohman, and Thomas McCarthy (eds.), *After Philosophy: End or Transformation.* Cambridge: The MIT Press.

Mackie, J.L., 1985. 'Locke and Representative Perception', in *Logic and Knowledge: Selected Papers*, vol. i. Oxford: Clarendon Press, pp. 214-224.

Madden, E.H. (ed.), 1960. *Theories of Scientific Method: The Renaissance through the Nineteenth Century.* Seattle: University of Washington Press.

Maddison, R.E.W., 1969. *The Life of the Honourable Robert Boyle, F.R.S.*. London: Taylor & Francis.

Mamiani, M., 1986. *Il Prisma del Newton: I Meccanismi dell'invenzione Scientifica*. Roma: Laterza.

—, 2000. 'The Structure of a Scientific Controversy. Hooke versus Newton about Colors', in P. Machamer, M. Pera, and A. Baltas (eds.), *Scientific Controversies. Philosophical and Historical Perspectives.* Oxford: Oxford University Press, pp. 143-52.

Marci, M., 1968. *Thaumantias, Liber de arcu coelesti deque colorum apparentium natura, ortu et causis,* ed. J. Marek. Prague: Pragopress.

Marion, J.L., 1998. 'The Idea of God', in D. Garber and M. Ayers (eds.), *The Cambridge History of Seventeenth-Century Philosophy.* Cambridge: Cambridge University Press, pp. 265-304.

Markley, R., 1999. 'Newton, Corruption, and the Tradition of Universal History', in J.E. Force and R.H. Popkin (eds.), *Newton and Religion. Context, Nature, and Influence.* Dordrecht: Kluwer Academic Publishers, pp. 121-43.

Marshall, J., 1992. 'John Locke and Latitudinarianism', in Richard Kroll, Richard Ashcraft, and P. Zagorin (ed.), *Philosophy, Science, and Religion in England 1640-1700.* Cambridge: Cambridge University Press, pp. 253-82.

—, 1994. *John Locke. Resistance, Religion, and Responsibility.* Cambridge: Cambridge University Press.

Martin, J., 1992. *Francis Bacon, the State, and the Reform of Natural Philosophy.* Cambridge: Cambridge University Press.

McCann, E., 1994. 'Locke's Philosophy of Body', in V. Chappell (ed.), *The Cambridge Companion to Locke.* Cambridge: Cambridge University Press.

McEvoy, J., 1979. 'Electricity, Knowledge and the Nature of Progress in Priestley's Thought', *British Journal for the History of Science,* 12: 1-30.

McGrew, T., 1992 'Unraveling Innate Ideas', *History of Philosophy Quarterly*, 9: 307-317.

McGuire, J.E., 1968. 'Force, Active Principles and Newton's Invisible Realm', *Ambix* 15: 154-208.

—, 1972. 'Boyle's Conception of Nature', *Journal of the History of Ideas*, 33: 523-42.

McGuire, J.E., and M. Tamny (eds), 1983. *Certain Philosophical Questions: Newton's Trinity Notebook.* Cambridge: Cambridge University Press.

McKendrick, N., J. Brewer, J.H. Plumb, 1982. *The Birth of a Consumer Society.* Bloomington: Indiana University Press.

McMullin, E., 1990. 'Conceptions of Science in the Scientific Revolution', in D.C. Lindberg and R. Westman (eds.), *Reappraisals of the Scientific Revolution.* Cambridge: Cambridge University Press, pp. 27-92.

—, 1998. 'Galileo on Science and Scripture', in P. Machamer, *The Cambridge Companion to Galileo.* Cambridge: Cambridge University Press, pp. 271-347.

Mehta, U.S., 1992. *The Anxiety of Freedom: Imagination and Individuality in Locke's Political Thought.* Ithaca: Cornell University Press.

Menn, S., 1998. 'The Intellectual Setting', in D. Garber and M. Ayers (eds.), *The Cambridge History of Seventeenth-Century Philosophy.* Cambridge: Cambridge University Press, pp. 33-86.

Merton, R.K., 1973. *The Sociology of Science. Theoretical and Empirical Investigations*, ed. N.W. Storer. Chicago: The University of Chicago Press.

Meynell, G., 1995. 'Locke, Boyle and Peter Stahl', *Notes and Records of the Royal Society*, 49: 185-92.

Michael, F.S., 1997. 'Why Logic Became Epistemology: Gassendi, Port Royal and the Reformation in Logic', in P.A. Easton, (ed.), *Logic and the Workings of the Mind: The Logic of Ideas and Faculty Psychology in Early Modern Philosophy.* Atascadero: Ridgeview Publishing.

Michael, F.S. and E. Michael, 1990. 'The Theory of Ideas in Gassendi and Locke', *Journal of the History of Ideas*, 51: 379-99.

Milton, J.R., 1994 'Locke's Life and Times', in V. Chappell (ed.), *The Cambridge Companion to Locke.* Cambridge: Cambridge University Press, pp. 5-25.

—, 1998. 'Laws of Nature', in D. Garber and M. Ayers (eds.), *The Cambridge History of Seventeenth-Century Philosophy.* Cambridge: Cambridge University Press, pp. 680-701.

Morgan, J., 1986. *Godly Learning.* Cambridge: Cambridge University Press.

Mortimer, C., 1745. *An Address to the Publick; containing narratives of the effects of certain chemical remedies in most diseases, etc.* London.

Mulligan, L., 1994. 'Robert Boyle, 'Right Reason', and the Meaning of Metaphor', *Journal of the History of Ideas*, 45: 235-257.

Murdin, L., 1985. *Under Newton's Shadow: Astronomical Practices in the Seventeenth Century.* Bristol: Adam Higer.

Nadler, S., 1998. 'Doctrines of Explanation in Late Scholasticism and in the Mechanical Philosophy', in D. Garber and M. Ayers (eds.), *The Cambridge History of Seventeenth-Century Philosophy.* Cambridge: Cambridge University Press, pp. 513-52.

Nakajima, H., 1984. 'Two Kinds of Modification Theory of Light: Some New Observations on the Newton-Hooke Controversy of 1672 Concerning the Nature of Light', *Annals of Science*, 41: 261-78.

Nelson, J.S., A. Megill, and D.N. McCloskey (eds.), 1987. *The Rhetoric of the Human Sciences: Language and Argument in Scholarship and Public Affairs.* Madison: University of Wisconsin Press.

Newman, W.R., 1994. *Gehennical Fire. The Lives of George Starkey, an American Alchemist in the Scientific Revolution.* Cambridge: Harvard University Press.

—, 1996. 'The Alchemical Sources of Robert Boyle's Corpuscular Philosophy', *Annals of Science*, 53: 567-85.

Newton, I., 1672. 'A letter of Mr. Isaac Newton…containing his new theory about light and colours…', *Philosophical Transactions* 6: 3075-87.

—, 1950. *Sir Isaac Newton's Theological Manuscripts,* ed. H. Mclachlan. Manchester: Manchester University Press.

—, 1952. *Opticks or A Treatise of the Reflections, Refractions, Inflections & Colours of Light.* New York: Dover Publications.

—, 1959-1977. *The Correspondence of Isaac Newton*, ed. H.W. Turnbull, J.F. Scott, A.R. Hall and L. Tilling, 7 vols. Cambridge: Cambridge University Press.

—, 1967-1981. *The Mathematical Papers of Isaac Newton*, ed. D.T. Whiteside, 8 vols. Cambridge: Cambridge University press.

—, 1974. 'Treatise on revelation', in Frank E. Manuel, *The Religion of Isaac Newton.* Oxford: Clarendon Press.

—, 1984. *The Optical Papers of Isaac Newton*, ed. A.E. Shapiro. Cambridge: Cambridge University Press.

—, 1999. *Isaac Newton, The Principia: Mathematical Principles of Natural Philosophy; a new translation by I. Bernard Cohen and Anne Whitman, assisted by Julia Budenz; preceded by a guide to Newton's Principia by I. Bernard Cohen*. Berkeley: University of California Press.

Newton Harvey, E., 1957. *A History of Luminescence. From the earliest Times Until 1900.* Philadelphia: American Philosophical Society.

Nicolson, M., 1929-30. 'Christ's College and the Latitude-Men', *Modern Philology*, 27: 35-53.

Nollet, J.A., 1746. *Essai sur L'électricité des corps*. Paris.

North, D.C., 1991. 'Institutions', *Journal of Economic Perspectives*, 5: 97-112.

Norton, D.F., 1980. 'The Myth of 'British Empiricism', *History of European Ideas*, 1: 331-44.

Nuchelmans, G., 1998. 'Proposition and Judgement', in D. Garber and M. Ayers (eds.), *The Cambridge History of Seventeenth-Century Philosophy.* Cambridge: Cambridge University Press, pp. 118-131.

Nussbaum, M.C. and A. Oksenberg Rorty (eds.), 1995. *Essays on Aristotle's De anima*. Oxford: Clarendon Press.

Oakley, F. and E.W. Urdang 1991. 'Locke, Natural Law, and God', in R. Ashcraft (ed.), *John Locke Critical Assessments,* 4 vols. London: Routlege, ii, pp. 63-83.

Oldenburg, H., 1965-1986. *The Correspondence of Henry Oldenburg,* ed. A.R. Hall and M.B. Hall, 13 vols. Madison: University of Wisconsin Press.

Osler, M.J., 1970. 'John Locke and the Changing Ideal of Scientific Knowledge', *Journal of the History of Ideas*, 31: 3-16.

—, 1992. 'The Intellectual Sources of Boyle's Philosophy of Nature: Gassendi's Voluntarism and Boyle's Physico-Theological Project', in R. Kroll, R. Ashcraft, and P. Zagorin (eds.), *Philosophy, science and religion in England 1640-1700.* Cambridge: Cambridge University Press, pp. 178-198.

—, 1999. 'Mixing Metaphors: Science and Religion or Natural Philosophy and Theology in Early Modern Europe', *History of Science*, 35: 91-113.

— (ed.), 2000. *Rethinking the Scientific Revolution.* Cambridge: Cambridge University Press.

Oster, M., 1992. 'The Scholar and the Craftsman Revisited: Robert Boyle as Aristocrat and Artisan', *Annals of Science*, 49: 255-76.

—, 1993. 'Biography, Culture, and Science: The Formative Years of Robert Boyle', *History of Science*, 31: 177-226.

—, 1994. 'Virtue, Providence and Political Neutralism: Boyle and Interregnum Politics', in M. Hunter (ed.), *Robert Boyle Reconsidered.* Cambridge: Cambridge University Press, pp 19-36.

O'Toole, R.J.,1974. 'Qualities and Powers in the Corpuscular Philosophy of Robert Boyle', *Journal of the History of Philosophy*, 12: 295-315.

Park, K., 1988. 'The Organic Soul', in C.B. Schmitt *et al.* (eds.), *The Cambridge History of Renaissance Philosophy.* Cambridge: Cambridge University Press, pp. 464-84.

Parker, S., 1662. *A Brief Account of the new Sect of Latitude-Men.* Cambridge.

Pasnau, R., 1997. *Theories of Cognition in the Later Middle Ages.* Cambridge University Press.

Passmore, J.A., 2000. 'Locke and the Ethics of Belief', in G. Fuller, R. Stecker, J.P. Wright (eds.), *John Locke.* An Essay Concerning Human Understanding *in focus.* London: Routledge, pp. 187-209.

Patrick, S., 1669. *A Friendly Debate Between a Conformist and a Non-Conformist.* London.

Pérez-Ramos, A., 1988. *Francis Bacon's Idea of Science and the Maker's Knowledge Tradition.* Oxford: Clarendon Press.

Perkins, W., 1609. *The Works*, 2vols. London.

Peterfreund S., (ed.), 1990. *Literature and Science: Theory and Practice.* Boston: Northeastern University Press.

Pickering, A. (ed.), 1992. *Science as Culture and Practice.* Chicago: University of Chicago Press.

—, 1995. *The Mangle of Practice: Time, Agency, and Science.* Chicago: The University of Chicago Press.

—, 1995a. 'Beyond Constraints: The Temporality of Practice and the Historicity of Knowledge', in J.Z. Buchwald (ed.), *Scientific Practice. Theories and Stories of Doing Physics.* Chicago: The University of Chicago Press, pp. 42-55.

Pitt, J.C. (ed.), 1988. *Theories of Explanation.* Oxford: Oxford University Press.

Plumb, J.H., 1972. 'The Public, Literature and the Arts in the 18th Century', in P. Fritz and D. Williams (eds.), *The Triumph of Culture: 18th Century Perspectives*, Toronto: A.M. Hakkert.

—, 1973. 'Reason and Unreason in the Eighteenth Century: the English Experience', in *In the Light of History.* London: Allen Lane.

Popkin, R.H., 1953. 'Joseph Glanvill: A Precursor of David Hume', *Journal of the History of Ideas*, 14: 292-303.

—, 1979. *The History of Scepticism from Erasmus to Spinoza.* Berkeley: University of California Press.

—, 1988. 'Theories of Knowledge', in C.B. Schmitt *et al.* (eds.), *The Cambridge History of Renaissance Philosophy.* Cambridge: Cambridge University Press, pp. 668-84.

Popper, C., 1968. *The Logic of Scientific Discovery.* London: Routledge & Kegan Paul.

Porter, R., 1980. 'Science, Provincial Culture and Public Opinion in Enlightenment England', *British Journal for Eighteenth-Century Studies*, 3: 20-46.

—, 1981. 'The Enlightenment in England', in R. Porter and M. Teich (eds.), *The Enlightenment in National Context.* Cambridge: Cambridge University Press.

Porter R., and M. Teich, 1992. *The Scientific Revolution in National Context.* Cambridge: Cambridge University Press.

Priestley, J., 1772. *The History and Present State of Discoveries Relating to Vision, Light, and Colours,* 2 vols. London.

—, 1775. *The History and Present State of Electricity, with Original Experiments*, 2 vol. London.

Principe, L.M., 1994. 'Boyle's Alchemical Pursuits', in M. Hunter (ed.), *Robert Boyle Reconsidered.* Cambridge: Cambridge University Press, pp. 91-105.

—, 1994a. 'Style and Thought of the Early Boyle. Discovery of the 1648 Manuscript of *Seraphic love*', *Isis*, 85: 247-60.

—, 1995. 'Virtuous Romance and Romantic Virtuoso: The Shaping of Robert Boyle's Literary Style', *Journal of the History of Ideas*, 56: 377-97.

—, 1998. *The Aspiring Adept: Robert Boyle and his Alchemical Quest.* Princeton: Princeton University Press.

—, 2000. 'The Alchemies of Robert Boyle and Isaac Newton: Alternate Approaches and Divergent Developments', in M.J. Osler (ed.), *Rethinking the Scientific Revolution*, Cambridge: Cambridge University Press, pp. 201-20.

Pumfrey, S., 1987. 'Mechanizing Magnetism in Restoration England: The Decline of Magnetic Philosophy', *Annals of Science*, 44: 1-21.

Rabb, T.K., 1969. 'Francis Bacon and the Reform of Society', in T.K. Rabb and Jerrold E. Seigel (eds.), *Action and Conviction in Early Modern Europe. Essays in Memory of E.H. Harbison.* Princeton: Princeton University Press, pp. 169-93.

Ramati, E., 2001. 'The Hidden Truth of Creation: Newton's Method of Fluxions', *British Journal for the History of Science,* 34: 417-38.

Randall, J.H., 1940. 'The Development of Scientific Method in the School of Padua', *Journal of the History of Ideas*, 1:177-206.

—, 1961. *The School of Padua and the Emergence of Modern Science.* Padova: Editrice Antenore.

Rattansi, P.M., 1968. 'The Intellectual Origins of the Royal Society', *Notes and Records of the Royal Society*, 23:129-43.

Rattansi, P., and A. Clericuzio (eds.), 1994. *Alchemy and Chemistry in the Sixteenth and Seventeenth Centuries.* Dordrecht: Kluwer.

Reeds, K.M., 1991. *Botany in Medieval and Renaissance Universities.* New York: Garland.

Reif, P., 1969. 'The Textbook Tradition in Natural Philosophy,1600-1650', *Journal of the History of Ideas*, 30: 17-32.

Ribe, N.M., 1997. 'Cartesian Optics and the Mastery of Nature', *Isis*, 88: 42-61.

Richards, J., L. Mulligan, and J.K. Graham, 1981. ''Property' and 'People': Political Usages of Locke and Some Contemporaries', *Journal of the History of Ideas*, 42: 29-51.

Rogers, G.A.J., 1978. 'Locke's *Essay* and Newton's *Principia*', *Journal of the History of Ideas*, 39: 217-32.

Rogers, P., 1974. *The Augustan Vision*, London: Weidenfeld & Nicolson.

—, 1985. *Literature and Popular Culture in Eighteenth Century England.* Sussex: Harvester Press.

Roller, D. and D.H.D. Roller, 1954. 'The Development of the Concept of Electric Charge', J.B. Conant, ed. *Harvard Case Histories in Experimental Science*, Cambridge, MA.

Ronchi, V., 1970. *The Nature of Light. An Historical Survey,* trans. V. Barocas. Cambridge: Harvard University Press.

Rowbottom, M. E., 1965. 'John Theophilus Desaguliers (1683-1744)', *Huguenot Society Proceedings*, 21: 196-218.

Rudavsky, T.M., 2001. 'Galileo and Spinoza: Heroes, Heretics, and Hermeneutics', *Journal of the History of Ideas*, 62: 611-31.

Rudwick, M.J.S., 1985. *The Great Devonian Controversy: the Shaping of Scientific Knowledge among Gentlemanly Specialists.* Chicago: The University of Chicago Press.

Sabra, A.I., 1967. *Theories of Light from Descartes to Newton.* London: Oldbourne.

Sailor, D.B., 1988. 'Newton's Debt to Cudworth', *Journal of the History of Ideas*, 49: 511-18.

Sarasohn, L.T., 1996. *Gassendi's Ethics. Freedom in a Mechanistic Universe.* Ithaca: Cornell University Press.

Sargent, R.M., 1986. 'Robert Boyle's Baconian Inheritance', *Studies in History and Philosophy of Science*, 17: 469-86.

—–, 1995. *The Diffident Naturalist: Robert Boyle and the Philosophy of Experiment*. Chicago: The University of Chicago Press.

Schaffer, S., 1980. 'Natural Philosophy', in G.S. Rousseau and R. Porter (eds.), *The Ferment of Knowledge.* Cambridge: Cambridge University Press.

—–, 1983. 'Natural Philosophy and Public Spectacle in the Eighteenth Century', *History of Science*, 21: 1-43.

—–, 1985. 'Occultism and Reason', in A.J. Holland (ed.), *Philosophy, its History and Historiography*. Dordrecht: D. Reidel, pp. 117-143.

—–, 1987. 'Godly Men and Mechanical Philosophers: Souls and Spirits in Restoration Natural Philosophy', *Science in Context,* 1: 55-85.

—–, 1989. 'Glass Works: Newton's Prisms and the Uses of Experiment', in D. Gooding, T. Pinch, S. Schaffer (eds.), *The Uses of Experiment.* Cambridge: Cambridge University Press, pp. 77-104.

—–, 1997. 'Experimenters' Techniques, Dyers' Hands, and the Electrical Planetarium', *Isis*, 88: 456-83.

Schofield, R.E., 1953. 'John Wesley and Science in Eighteenth Century England', *Isis*, 44: 331-40.

Seaver, P.S., 1970. *The Puritan Lectureships: The Politics of Religious Dissent 1560-1662.* Stanford: Stanford University Press.

Shadwell, T., 1966. *The Virtuoso*, ed. M.H. Nicolson and D.S. Rodes. Lincoln: University of Nebraska Press.

Shanahan, T., 1988. 'God and Nature in the Thought of Robert Boyle.' *Journal of the History of Philosophy*, 26: 547-69.

—, 1994. 'Teleological Reasoning in Boyle's *Disquisitions about Final Causes*', in M. Hunter (ed.), *Robert Boyle Reconsidered.* Cambridge: Cambridge University Press, pp. 177-92.

Shapin, S., 1979. 'Homo Phrenologicus: Anthropological Perspectives on an Historical Problem', in B. Barnes and S. Shapin (eds.), *Natural Order.* Beverly Hills and London: Sage.

—, 1981. 'Of Gods and Kings: Natural Philosophy and Politics in the Leibniz-Clarke Disputes', *Isis*, 72: 187-215.

—, 1982. 'History of Science and its Sociological Reconstructions', *History of Science*, 20: 157-211.

—, 1988. 'The House of Experiment in Seventeenth-Century England', *Isis*, 79: 373-404.

—, 1988a. 'Robert Boyle and Mathematics: Reality, Representation, and Experimental Practice,' *Science in Context*, 2: 23-58.

—, 1994. *A Social History of Truth. Civility and Science in Seventeenth-Century England.* Chicago: The University of Chicago Press.

—, 1996. *The Scientific Revolution.* Chicago: The University of Chicago Press.

Shapin, S. and S. Schaffer, 1985. *Leviathan and the Air-Pump: Hobbes, Boyle, and the Experimental Life.* Princeton: Princeton University Press.

Shapiro, A.E., 1973. 'Kinematic Optics: a Study of the Wave Theory of Light in the Seventeenth Century', *Archive for History of Exact Sciences*, 11: 134-266.

—, 1980. 'The Evolving Structure of Newton's Theory of White Light and Color', *Isis*, 71: 211-35.

—, 1990. 'The *Optical Lectures* and the Foundations of the Theory of Optical Imagery', in M. Feingold (ed.), *Before Newton. The life and times of Isaac Barrow.* Cambridge: Cambridge University Press, pp. 105-78.

—, 1996. 'The Gradual Acceptance of Newton's Theory of Light and Colors, 1672-1727', *Perspectives on Science*, 4: 59-140.

Shapiro, B.J., 1969. *John Wilkins 1614-1672. An Intellectual Biography.* Berkeley: University of California Press.

—, 1974. 'Latitudinarianism and Science in Seventeenth-Century England', in C. Webster (ed.), *The Intellectual Revolution of the Seventeenth Century,* Past and Present Series, London, pp. 286-316.

—, 1983. *Probability and Certainty in Seventeenth-Century England.* Princeton: Princeton University Press.

—, 2000. *A Culture of Fact. England, 1550-1720.* Ithaca: Cornell University Press.

Shea, W.R., 1991. *The Magic of Numbers and Motion.* Canton: Science History Publications.

Simon, D.L. and P.L. Hinkley, 2001. 'David Gregory on Newton's Burning Mirror', *Notes and Records of the Royal Society*, 55: 185-190.

Simon, G., 1975. 'On the Theory of Visual Perception of Kepler and Descartes: Reflections on the Role of Mechanism in the Birth of Modern Science', in A. Beer and P. Beer (eds.), *Kepler: Four Hundred Years,* (*Vistas in Astronomy,* 18), pp. 825-32.

Simon, I., 1967. *Three Restoration Divines: Barrow, South, Tillotson.* Paris: Société d'edition "Les Belles Lettress".

Simon, J., 1966. *Education and Society in Tudor England.* Cambridge: Cambridge University Press.

Slaughter, M.M., 1982. *Universal Language and Scientific Taxonomy in the Seventeenth Century.* Cambridge: Cambridge University Press.

Smith, A.M., 1999. *Ptolemy And the Foundations of Ancient Mathematical Optics: A Source Based Guided Study.* Philadelphia: American Philosophical Society.

Snobelen, S.T., 2001. ''God of Gods, and Lord of Lords':' The Theology of Isaac Newton's General Scholium to the *Principia*', *Osiris* 16: 169-208.

Sprat, T., 1667. *The History of the Royal-Society of London for the improving of natural knowledge*, London.

—, 1696. *A discourse made by the Ld Bishop of Rochester to the clergy of his diocese, at his visitation in the year 1695.* London.

Spurr, J., 1991. *The Restoration Church of England, 1646-1689.* New Haven: Yale University Press.

Stewart, L., 1986. 'Public Lectures and Private Patronage in Newtonian England', *Isis*, 77: 47-58.

—, 1992. *The Rise of Public Science. Rhetoric, Technology, and Natural Philosophy in Newtonian Britain, 1660-1750.* Cambridge: Cambridge University Press.

Stewart. M.A., 1981 'Locke's Professional Contacts with Robert Boyle', *The Locke Newsletter*, 12: 19-44.

Stocker, M., 1992. 'From Faith to Faith in Reason? Religious Thought in the Seventeenth Century', in J.G.S. Cain and K. Robinson (eds.), *Into Another Mould: Change and Continuity in English Culture, 1625-1700.* London: Routledge.

Stone, L., 1980 'The Results of the English Revolutions of the Seventeenth Century', in J.G.A. Pocock (ed.), *Three British Revolutions: 1641, 1688, 1776.* Princeton: Princeton University Press, pp. 23-108.

Stukeley, W., 1883. 'The Family Memoirs of the Rev. William Stukeley, M.D.', *Publications of the Surtees Society* 76, London.

Taylor, C., 1987. 'Beyond Epistemology', in K. Baynes, J. Bohman, T McCarthy (eds.), *After Philosophy: End or Transformation?* Cambridge: MIT Press.

Tipton, I.C., 1983. *Locke - Reason and Experience.* Milton Keynes: Open University Press.

—, 1996 'Locke: Knowledge and its Limits', in S. Brown (ed.), *British Philosophy and the Age of Enlightenment.* London: Routledge.

Todd, M., 1987. *Christian Humanism and the Puritan Social Order.* Cambridge: Cambridge University Press.

Tuck, R., 1979. *Natural Rights Theories. Their Origin and Development.* Cambridge: Cambridge University Press.

Tully, J., 1988 'Governing Conduct', in E. Leites (ed.), *Conscience and Casuistry in Early Modern Europe.* Cambridge: Cambridge University Press, pp. 12-71.

van Fraassen, B.C., 1980. *The Scientific Image.* Oxford: Oxford University Press.

—, 1985. 'Empiricism in the Philosophy of Science', in P.M. Churchland, and C.A. Hooker (eds.), *Images of Science. Essays on Realism and Empiricism, with a Reply from Bas C. van Fraassen.* Chicago: The University of Chicago Press, pp. 245-308.

van Helden, A., 1983. 'The Birth of the Modern Scientific Instrument', in J.G. Burke (ed.), *The Uses of Science in the Age of Newton.* Berkeley: University of California Press.

—, 1994. 'Telescopes and Authority from Galileo to Cassini', *Osiris*, 9: 9-29.

van Leeuwen, H.G., 1963. *The Problem of Certainty in English Thought.* The Hague: Nijhoff.

Wallace, W.A., 1959. *The Scientific Methodology of Theodoric of Freiberg.* Fribourg: The University Press.

Ward, S., 1654. *Vindiciae Academiarum.* Oxford.

Watson, W., 1744-5. 'Experiments and Observations tending to illustrate the Nature and Properties of Electricity', *Philosophical Transactions*, 43: 481-501.

—, 1746. 'Further Experiments and Observations', *Philosophical Transactions*, 44: 41-50.

Watt, I., 1967. *The Rise of the Novel: Studies in Defoe, Richardson and Fielding.* Harmondsworth: Pelican.

Webster, C., 1965. 'The Discovery of Boyle's Law, and the Concept of the Elasticity of Air in the Seventeenth Century', *Archive for History of Exact Sciences*, 2: 441-502.

—, 1975. *The Great Instauration: Science, Medicine, and Reform 1626-1660.* London: Duckworth.

Westfall, R.S., 1956. 'Unpublished Boyle Papers Relating to Scientific Method', *Annals of Science*, 12:63-73, 103-17.

—, 1962a. 'The Development of Newton's Theory of Color', *Isis*, 53: 339-58.

—, 1962b. 'Newton and his Critics on the Nature of Colours', *Archives internationales d'histoire des sciences*, 15: 47-58.

—, 1963. 'Newton's Reply to Hooke and the Theory of Colours', *Isis*, 54: 82-96.

—, 1966. 'Newton Defends his First Publication: the Newton-Lucas Correspondence', *Isis*, 57: 299-314.

—, 1980. *Never at Rest. A Biography of Isaac Newton.* Cambridge: Cambridge University Press.

Westman R.S., 1980. 'The Astronomer's Role in the Sixteenth Century: A Preliminary Study', *History of Science*, 18: 104-47.

Wheler, G., 1739-40. 'Some Electrical Experiments, chiefly regarding the Repulsive Force of Electrical Bodies', *Philosophical Transactions*, 41: 98-117.

Wickes, H., 1994. 'Descartes's Denial of the Autonomy of Reason', in J. Cottingham, (ed.), *Reason, Will, and Sensation. Studies in Descartes's Metaphysics.* Oxford: Clarendon Press.

Wilkins, J., 1669. *Ecclesiastes: or, the Gift of Preaching as it falls under the rules of art.* London.

—, 1678. *Of the Principles and Duties of Natural Religion: Two Books.* London.

Williams, B., 1978 *Descartes: The Project of Pure Enquiry.* Hassocks: Harvester Press.

Willis, T., 1684. 'Of Fermentation', in *Practice of Physick,* trans. S. Pordage. London.

Wise, M.N. and C. Smith, 1986. 'Measurement, Work, and Industry in Lord Kelvin's Britain', *Historical Studies in the Physical Sciences*, 17: 147-73.

Wittgenstein, L., 1953. *Philosophical Investigations*, trans. G.E.M. Anscombe. New York: Macmillan.

Wojcik, J.W., 1997. *Robert Boyle and the Limits of Reason.* Cambridge: Cambridge University Press.

—, 2000. 'Pursuing knowledge: Robert Boyle and Isaac Newton', in M.J. Osler (ed.), *Rethinking the Scientific Revolution*, Cambridge: Cambridge University Press, pp. 183-200.

Wolterstroff, N., 1996. *John Locke and the Ethics of Belief.* Cambridge: Cambridge University Press.

Woolgar, S., 1988. *Science: The Very Idea.* London: Tavistock Publications.

Woolhouse, R.S., 1971. *Locke's Philosophy of Science and Knowledge.* Oxford: Blackwell.

—, 1994 'Locke's Theory of Knowledge', in V. Chappell (ed.), *The Cambridge Companion to Locke.* Cambridge: Cambridge University Press, pp. 146-171.

Yates, F., 1964. *Giordano Bruno and the Hermetic Tradition.* London: Routledge & Kegan Paul.

Yolton, J., 1956. *Locke and the Way of Ideas.* Oxford: Oxford University Press.

—, 1970. *Locke and the Compass of Human Understanding: A Selective Commentary on the Essay.* Cambridge: Cambridge University Press.

—, 1975. 'Ideas and Knowledge in Seventeenth-Century Philosophy', *Journal of the History of Philosophy*, 13: 145-165.

—, 1975a. 'On Being Present to the Mind: A Sketch for the History of an Idea', *Dialogue*, 14: 373-88.

—, 1991. 'Locke on the Law of Nature', in R. Ashcraft (ed.), *John Locke Critical Assessments,* 4 vols. London: Routlege, ii, .pp. 16-33.

Young, R.F., 1971. *Comenius in England.* New York: Arno Press.

Zagorin, P., 2001. 'Francis Bacon's Concept of Objectivity and the Idols of the Mind', *British Journal for the History of Science*, 34: 379-393.

Zilsel, E., 1942. 'The Sociological Roots of Science', *American Journal of Sociology*, 47: 544-62.

—, 1945. 'The Genesis of the Concept of Scientific Progress', *Journal of the History of Ideas*, 6: 325-49.

Index